Stephanos Theodoros Xenos

Depredations - Overend, Gurney, & Co

And the Greek and Oriental steam navigation company

Stephanos Theodoros Xenos

Depredations - Overend, Gurney, & Co
And the Greek and Oriental steam navigation company

ISBN/EAN: 9783743467149

Manufactured in Europe, USA, Canada, Australia, Japa

Cover: Foto ©berggeist007 / pixelio.de

Manufactured and distributed by brebook publishing software (www.brebook.com)

Stephanos Theodoros Xenos

Depredations - Overend, Gurney, & Co

DEPREDATIONS:

OVEREND, GURNEY, & CO.,

AND THE

GREEK & ORIENTAL STEAM NAVIGATION COMPANY.

BY

STEFANOS XENOS.

London:
PUBLISHED BY THE AUTHOR,
AT
No. 9, ESSEX STREET, STRAND.

1869.

PREFACE.

WHOEVER may attentively peruse this little volume will perceive that it was impossible for me to enter into an explanation of my transactions with Messrs. Overend, Gurney, and Co. without first giving some account of the formation of the Greek and Oriental Steam Navigation Company, and its existence for three years previously to the period at which I entered into relations with the great Lombard Street capitalists. Such account was indispensable to the clear understanding of the after-narrative.

With regard to the conversations recorded in this volume, the accuracy of which may be questioned by the sceptical, I ought, perhaps, to offer some explanation. In the first place, I have, during many years, been in the habit of taking notes and keeping a diary of all important transactions in which I have been engaged. Secondly I never destroy letters or documents; so that, in case of necessity, I have authorities to which I can refer, and, if need be, produce. And, thirdly, as far as I am personally concerned, I can always rely on my memory, which is singularly clear and retentive, particularly—though possibly I am not singular in this respect—in cases where I have been wronged and ill-treated. I can then recall correctly, not alone the words of my opponents, but every look, and movement, and gesture that have aggravated the wrong-doing by which I suffered. Of the conversations reported here, I can conscientiously say they are substantially what the speakers expressed; and they would be *literally* the same, but that many of these conversations were originally uttered in Greek, and not a few in broken English. It was, therefore, necessary that the one should be translated and the others corrected before going to press.

Despite the care with which I have always guarded my business documents, I had, in 1865, the mortification to lose a case containing many important papers, amongst others, Mr. Edwards's letter in reference to the yacht, and the original of the printed circular to which I refer at page 217. It was in removing some of my furniture from Petersham Lodge that this mischance occurred. A diligent but fruitless search was made when the loss was ascertained. (See Appendix, No. 24.) The absence of some of these documents has not alone delayed the publication of this volume, but obliged me to suppress all allusion to many important points, about which I was quite clear, but would not mention without being in a position to produce corroborative proofs.

In further authentication of the correctness of that which is set down in the following pages, it may not be amiss to mention that this history of the Greek and Oriental Steam Navigation Company, which I find myself obliged to hurry into publicity, formed originally portion of a work which was not intended to see the light for many years. It was written whilst the facts were fresh in my memory—whilst the words still vibrated in my ears.

In publishing this little volume, I feel that I am making my defence at the bar of public opinion. I hope my auditory will remember that I was put upon my trial without notice or warning. Should it be urged against me that I make free with the names of other men, let it not be forgotten that my name was not spared. What I say is spoken solely in self-defence. With the materials at my command, I might have sent forth to the world a sensational volume. Had I sought a *succès de salon*, I might have garnished my narrative with savoury morsels of scandal, with piquant allusions to many an operatic Zephyrina and dramatic Eucharis; I might have told tales of Pretty Horsebreakers and capricious Anonymas; or I might have hinted at certain titled Calypsos, at whose bidding some of our good City men entered the commercial lists, and there fought the most desperate combats. And I might, indeed, on the other hand, have talked of many patient Griseldas, who, whilst their husbands lavished thousands on worthless rivals, stinted themselves in martyrizing economy, trying to counterbalance reckless extravagance. With all these accessories I have dispensed, contenting myself with an unvarnished narrative of dry facts. Were the century a few years older, had time toned down the memory of the present, I might, in a more voluminous work, have laid before the public more realistic views of the great panorama which is daily being revolved in this GREAT PANANTHROPOPOLIS.

CONTENTS.

CHAP.		PAGE
I.	The First Chartered Steamers	1
II.	The First Ownership	8
III.	Profits and Contracts	14
IV.	The First Clouds	20
V.	The Galway Steam Company	26
VI.	A Commercial Perseus	34
VII.	My Agents	38
VIII.	A Grave Mistake	46
IX.	Disappointment	50
X.	The Ultramarine Powder	54
XI.	The First Mortgages	58
XII.	The Miniature Court of Louis XIV.	64
XIII.	A Perilous Position	72
XIV.	Messrs. Overend, Gurney, and Co.	79
XV.	My First Financer	84
XVI.	My Second Financer	91
XVII.	The "British Star"	100
XVIII.	The Penelope	105
XIX.	Deeper in the Water	113
XX.	My Third Financer	119
XXI.	How Eighty Thousand Pounds were Lost	125
XXII.	A Breach of Faith	131
XXIII.	Zachariah Pearson	140
XXIV.	Who is entitled to a Commission?	152
XXV.	My Fourth Financer	157
XXVI.	Remortgaging the Steamers	163
XXVII.	A Second Breach of Faith	169
XXVIII.	A Third Breach of Faith	174
XXIX.	Criminal or Not?	182

CHAP.		PAGE
XXX.	Politics	187
XXXI.	The Election of Prince Alfred	193
XXXII.	My Illness	198
XXXIII.	Release of Petersham Lodge	207
XXXIV.	The Insolvency of Messrs. Overend, Gurney, and Co.	210
XXXV.	The Storm	219
XXXVI.	A Challenge	227
XXXVII.	The Settlement	232
XXXVIII.	Idleness	236
XXXIX.	Announcing to Messrs. Overend, Gurney, and Co. their Insolvency	241
XL.	The Philhellenic Committee	247
XLI.	The Stock Exchange	255
XLII.	Messrs. Overend, Gurney, and Co. a Limited Company	262
XLIII.	Battle between the Jews and the Greeks . .	267
XLIV.	The Anglo-Greek Steam Navigation and Trading Company	273
XLV.	The Consulate	286
XLVI.	Depredations	293
XLVII.	The Rats and the Pig	295
XLVIII.	Conclusion	304
	THE CITY OF LONDON	307
	APPENDIX	337

CHAPTER I.

THE FIRST CHARTERED STEAMERS.

In the year 1856 I was running a line of sailing vessels from London to the Levant and Black Sea. The chief import and export trade of England with that part of the world is in the hands of the Greek houses of England; and I must say my line of packets enjoyed a marked and special patronage, and was yielding a lucrative return. Whilst I was running several small clippers to the Levant, Messrs. Smith, Sundius, and Co., with Mr. A. G. Robinson, were running steamers on the same line. These gentlemen had to contend with a brisk competition on the part of the Liverpool lines, as well as with some London steamers; but as the outward and homeward freights for steamers were, in those days, three times greater than at present, their profits must have been enormous.

Having maturely reflected on these facts, and seeing that the export trade between London and the Levant had nearly reached 7000 tons per month, I resolved to substitute steamers for my sailing packets. In entering into competition with the established steam owners, I determined to profit by the errors they were committing, and avoid them. Amongst these I shall mention two which were a cause of daily complaint. In the first place, merchants, after having announced their goods to their correspondents, were not always able to find room for them on board the steamers; and secondly, merchants who applied to the steam owners and agents within the last few days of the steamer's stay in London, were often obliged to pay 10*s.* or even £1 per ton higher freight than those who had shipped first. This latter custom had enabled the owners of the steamers to take a very high tone. Owing to these difficulties in shipping, there were incessant complaints from Constantinople, Smyrna, Odessa, and other Levantine towns. Disagreements were constantly occurring between London merchants and their foreign consignees, and in more than one instance commission agents had lost their appointments on suspicion of incapacity, or worse.

This condition of things, combined with a promise of efficient support on the part of the shippers, had naturally great weight in inducing me to turn my line of sailing packets into steamers. But to start a line of steamers requires a large capital. I applied to Messrs. Lascaridi and Co., of 32, Bucklersbury; that firm came to my assistance. The Messrs. Lascaridi had two steamers, the Aleppo and the Beyrout. Mr. George Lascaridi, whose name recurs frequently in these pages, was at that time the sole representative of the London house of Lascaridi and Co. This gentleman placed the above-named steamers in my hands. We ran them conjointly with the James Brown and the Britannia. The two last-named steamers entailed a small loss in the first voyage, owing to our having miscalculated their carrying capacity. The company thus started in May, 1857, I called "The Greek and Oriental Steam Navigation Company." As soon as the above-mentioned charters had expired, we time-chartered the Hercules, Milo, and Admiral

Kanaris, at 22s. 6d. per ton per month, belonging to Mr. Edward Gourley, of Sunderland, now M.P., and others. Thus the Greek and Oriental Steam Navigation Company sprang into existence in 1857[*].

The steam power of the company consisted, at first starting, of five vessels—the Beyrout, the Aleppo, the Hercules, the Milo, and the Admiral Kanaris. In addition to these we had several sailing vessels. We were able to keep up a brisk competition with Messrs. Smith, Sundius, and Co., and others, as we enjoyed the exclusive patronage of the Greek houses of London and Manchester.

It is a recognized commercial truth, that on a time-charter the charterer seldom gains. That there should be a loss in our case was an inevitable consequence of the position in which my partners and I were placed. Neither Mr. George Lascaridi nor I had any experience in the management of steam property. Our steamers were of small power, and consumed enormous quantities of coal; and, above all, unfortunately we had paid an extravagant price per ton for the charter of these steamers. Then there was no organization amongst our agents; abuses crept in, and we had no check on their expenses. Neither had we any control over the captains and engineers, who, with a view to benefiting the owners of the steamers, navigated slowly; so that, in spite of the splendid freights I had secured, and the great support of the shippers, we again lost several thousands of pounds.

Mr. George Lascaridi, having the large business of his own firm to attend to, threw on my shoulders

---

[*] The Hercules was chartered in 1857; the Milo and Admiral Kanaris in March, 1858.

all the work and the responsibility of the Greek and Oriental Steam Navigation Company. He supplied the capital, I, through my friends, found the cargoes; but Mr. Lascaridi requested me to keep the name of my capitalist secret, for fear of injuring the credit of his firm. Owing to the observance of this secrecy, a certain amount of mystery, in the eyes of the commercial world, hung about the Greek and Oriental Steam Navigation Company.

About the year 1857 I began to understand something of the nature and value of steam property. I was educated by the Greek Government in the military school of Evelpides. I had studied there eight years, destined to become an officer in the Royal Engineers or the Royal Artillery. Owing to these early studies, I found no difficulty in mastering a knowledge of the mechanical structure of steamers. A long acquaintance with military discipline made the inspection of the staff of the steamers a kind of pleasure to a man whose first love was not commerce, and whom the political aspect of his own country had driven to England to engage in the labyrinthian perplexities of trade.

Having gone closely into details, I discovered to my great confusion that, of the £50,000 which we had paid the charterers for freights, £20,000 would have gone as profit into our pockets had we been the owners of the steamers. I found that the entire expenses of a steamer of that description for each voyage was £3500; this deducted from £6000, the gross freight, left a net profit of £2500 per voyage. Each steamer made four voyages in the year, so that here was a net profit of £10,000 per annum on a maritime property that only cost £16,000. It was evident that by eighteen months' work a steamer would pay for herself. The important point was to find the cargoes. This I was in a position to do abundantly through my large connexions. My friends gave me the preference and the first refusal of their goods. I drew the attention of Mr. Lascaridi to all these points, and urged upon him the necessity of purchasing our own steamers. But he was discouraged by the losses we had sustained, and was besides overburdened with other large transactions, so that he seldom came to my office. In fact, there was no written contract between us, nor had we ever entered into a thoroughly good verbal understanding. Add to this, that the Greek and Oriental Steam Navigation Company was to expire at the expiration of the charter-parties of the three steamers—the Hercules, the Milo, and the Admiral Kanaris. The Beyrout and the Aleppo I had been compelled to take out of the line after a couple of voyages, as being too small for the service.

Such was the position of affairs when several brokers, seeing the extensive business of the office, the credit of the Greek and Oriental Steam Navigation Company, and the great support it received from the Greek merchants, offered to sell me steamers on long credit. I went to Liverpool in the month of March, 1858, and there bought on credit, for £22,000, the General Williams, of 1152 tons register and 160 horse-power. In payment I was to give the bills of the Greek and Oriental Steam Navigation Company. The General Williams was in trust for the orphans of the late owner, and the trustee, Mr. Panton, of Sunderland, and the builder of the steamer, Mr. A. Leslie, of Newcastle, who had an interest in the vessel, wished to have the bills endorsed with another name. I tried Mr. Lascaridi; he at first positively refused, but after several days' discussion he decided upon accepting instead of endorsing the bills, and he would do this only on condition that the steamer should be registered in his name or in that of his nominee. He would at the same time give me a letter to the effect that when the steamer should have cleared herself in my line, he would then transfer the half ownership to me. In other words, I was to work the steamer to pay the bills accepted by Messrs. Lascaridi and Co., before I should be put in possession of my half ownership.

Messrs. Lascaridi and Co. had at this time in their employment, as managing clerk, Mr. Nicholas Koressios, a Greek. It was this Mr. Koressios who, on the 2nd of April, 1858, drew up the letter in which Messrs. Lascaridi and Co. informed me that they had bought the General Williams for £21,250; in it also, to my surprise, they informed me that they intended giving Mr. Koressios a $\frac{12}{44}$ share in the General Williams. Enclosed in this letter was the draft of the reply that I was to write them. I refused to sign any reply, and declined to accept Mr. Koressios as co-owner. Consequently the General Williams took the line without any contract being drawn up between me and Messrs. Lascaridi and Co., or between me and Mr. George Lascaridi.

I twice loaded the General Williams with excellent cargoes. In her second voyage she foundered off Malta, leaving a profit by the insurance and freight of about £8000. These profits were reduced to £5000 only by a long account, furnished by the house of Lascaridi of Liverpool, for needless repairs and for stores put on board the General Williams before leaving Liverpool for London; there was also a large item for commission on the purchase of the steamer. I demurred to these charges, and said if any one was entitled to commission on the purchase of the steamer, it was I, who had negotiated and concluded the transaction, and yet I had not made any charge. This and other transactions tended to weaken the good understanding that had subsisted between Mr. George Lascaridi and me.

CHAPTER II.

THE FIRST OWNERSHIP.

IN chartering the Hercules, Milo, and Admiral Kanaris, the Greek and Oriental Steam Navigation Company had undertaken to supply these vessels with fuel; we had accordingly sent large deposits of coal to our agents at Constantinople. These agents were the house of Lascaridi, known in the Levant as Messrs. Fachri, Lascaridi, and Co. These gentlemen furnished accounts of stupendous magnitude for discharging and loading the steamers; in fact, it became evident, on a comparison with the expenses of the steamers of other owners, that our agents did not understand their business, and were defrauded by those they employed. The exorbitant items in these accounts constituted the principal losses incurred by the three chartered steamers. I protested against these charges, and held Messrs. Lascaridi and Co., of London, responsible for the proceedings of their house at Constantinople. Hereupon Mr. George Lascaridi coolly informed me that the members of the house at Constantinople were not his partners, that I must look to them for the accounts in question, and that I might give the agency to whom I pleased. I replied:

"Oh, no! Lascaridi; I hold you responsible, because, when I talked of recommending the steamers, you said distinctly, 'Send them to our house at Constantinople.' The world believes it to be your house. Had you told me at the time that you would not be held responsible for the acts of Messrs. Fachri, Lascaridi, and Co., because they were not your partners, I would have recommended the steamers to some more competent firm. Besides, how comes it, when these gentlemen write to me, they always say, 'Ours of London,' and so forth? In short, I shall not accept their drafts, especially now that they hold large deposits of our coal, and that the price of coal is gone up at Constantinople, so that the coal is worth much more than when I sent it out."

I was compelled to take this determined tone, because I had learned, upon inquiry, that Messrs. Fachri, Lascaridi, and Co. were partners, under certain conditions, with the London house of Lascaridi and Co., and the latter being a partner in the Greek and Oriental, it was only just that the losses of the latter company should be shared by all the partners. But, so far from acting in this fashion, Mr. George Lascaridi wished to throw the entire burden of the losses sustained by the Greek and Oriental on that company alone, whilst the profits were to be divided half and half with the house of Lascaridi and Co. The reasons he gave for this mode of proceeding were, that when Lascaridi and Co., of London, accepted bills for the Greek and Oriental Steam Navigation Company, his partners abroad knew nothing of the transaction, and that when he communicated with them on the subject, they refused to have anything to do with steamers; consequently, he alone was my partner in the Greek and Oriental. The conclusion drawn from these pretexts by Mr. George Lascaridi was, that the Greek and Oriental Steam Navigation Company was to bear undivided every loss it incurred, and was at the same time to be reckoned a debtor to the house of Lascaridi and Co. for money advanced or bills accepted, for the chartered steamers of the company.

When Mr. George Lascaridi and I first started the Greek and Oriental Steam Navigation Company, I proposed to him to draw up a regular contract between us. His clerk, Nicholas Koressios, prepared a preliminary draft, in which he inserted his own name as my partner. This draft contained the most absurd conditions. The so-called "'Υπόμνημα," or memorandum, was drawn up in Greek by Mr. Koressios, who at that time possessed immense influence over the mind of Mr. George Lascaridi. It divided the shares of the Greek and Oriental thus:—

$\frac{1}{5}$ or 20 shares	. .	Lascaridi and Co.
$\frac{1}{5}$ or 20 do.	. .	G. P. Lascaridi.
$\frac{2}{5}$ or 40 do.	. .	Stefanos Xenos.
$\frac{1}{5}$ or 20 do.	. .	Nicholas Koressios.
100 shares.		

According to the terms of this 'Υπόμνημα I was to be manager of the company, Mr. G. P. Lascaridi to be president, and Mr. Nicholas Koressios to be 'Εξεταστής (controller) and vice-president. I refused to enter into such a contract. A sharp correspondence was exchanged between me and Mr. Koressios, and I finally forbade that gentleman to enter my office. Events showed that my penetration was not at fault. Within a very few months Mr. Koressios received permission to retire from the service of Lascaridi and Co.

After that I was never able to persuade Messrs. Lascaridi and Co. to draw up a deed of partnership between them and me. Mr. George Lascaridi used to tell me that he had boundless confidence in my honour, and did not care for a written contract. I could not do less than respond to such expressions of trust, and, of course, ceased to press the matter.

Seeing that the charter-parties of the Hercules, Milo, and Admiral Kanaris were about to expire, and that the Greek and Oriental would be left with only one steamer, the General Williams, and the sailing packets, I started, in company with Mr. John Preston, the ship-broker, for West Hartlepool, to inspect a large new steamer that Messrs. John Pile and Co. had then in their building-yard. She had been offered to me by Mr. John Preston on long credit. I was to give in payment the acceptances of the Greek and Oriental Steam Navigation Company.

Whilst inspecting this large steamer, I observed in the West Hartlepool Dock a small steam barge of very shallow draught, whose engines were at that moment being tried. I asked to whom it belonged. Mr. Joseph Spence, one of the partners of John Pile and Co., who accompanied us, said it belonged to his firm, and was a river barge. It was for sale. A project at that moment shot through my brain. I said to myself, "If I could send three such steam barges up the Danube into the shallow waters, let us say as far as Calafat and Oltenitza, where the markets are in which the grain is purchased that is brought down to Galatz and Ibraïla in carts, I should be able to buy wheat and Indian corn at at least 5s. or 6s. less per quarter than they can be bought at Galatz or Ibraïla. I could tranship this grain on board the large steamers at Sulina for England."

Without losing a moment, I entered into negotiations. I would take the large steamer, which was 1017 tons register, and of 220 nominal horse-power. I would take the little steam barge, and, two similar to be built within five months. After some discussion, it was agreed that for the entire lot I should give £33,850. In payment, I was to give the acceptances of the Greek and Oriental Steam Navigation Company from three to thirty months, which was equivalent to $2\frac{1}{2}$ years' credit. The acceptances were to be drawn by Messrs. Lascaridi and Co. The steamers were to be delivered free of mortgage.

I did not finally close the contract, because, before doing so, it was necessary to speak with Mr. George Lascaridi. I immediately returned to town and saw him. I said:

"Now, George, you know very well that this company, instead of entailing losses, would have left splendid profits if we had had our own steamers. Instead of losing any more time disputing about the accounts, and trying to determine whether you or Messrs. Lascaridi are partners in the Greek and Oriental Company, let us, like sensible men, try to repair past losses, and organize the line thoroughly. You see I have the exclusive support of the Greek merchants in London. In the Levant I am popular, owing to my family name, my father's patriotic services, and the great sacrifices he has made for Greece. I may perhaps add, that I am favourably known through my literary works."

Lascaridi would have interrupted. "Listen," I said; "there is another point. If, instead of chartering our steamers, we can buy them, and load them on our own account with grain at 4s. per quarter less than other Greek houses are paying, surely, if these houses are making money, we ought to make a fortune. We could order our own steamers whither we liked, whilst a charterer could only send his cargo to be discharged at the port whither it is bound."

A long conversation ensued, in which I tried to induce him to enter into an agreement on behalf of his firm, with me, to purchase, on long credit, three large steamers and three small ones. I showed him clearly that the bills would be provided for before coming to maturity, by the earnings of the steamers.

He said he could not draw an agreement between his firm and me, but he was willing to make one between himself and me. I simply observed that he was not a great capitalist. He ultimately said that he would not accept any more bills in the name of his firm, but that, if I would consent to pay a commission to the firm, he would endorse my bills with the signature of Lascaridi and Co., leaving it to my generosity, should the Greek and Oriental Steam Navigation Company prosper, to do by him what was just and proper.

This was a noble offer. It is not often one meets a friend capable of acting so generously. I thanked him with all my heart; indeed, my gratitude was boundless. I now closed my contract with Messrs. John Pile and Co., and within a few days Lascaridi drew and endorsed in my office the acceptances of the Greek and Oriental Steam Navigation Company, on behalf of Messrs. Lascaridi and Co.

The steamers were registered in my name, free of any kind of mortgage. I named the large steamer Admiral Miaoulis; the steam barge I called Bobolina. Those that Messrs. John Pile and Co. had commenced to build for me I named respectively the Botassis and the Tzamados.

CHAPTER III.

PROFITS AND CONTRACTS.

From the date of the above-mentioned contract, the business of the Greek and Oriental Steam Navigation Company began to flourish, and soon yielded splendid profits. Fortune smiled on all my undertakings. The Admiral Miaoulis, a magnificent vessel, I bought for £5000 less than her real value. The building of this steamer was commenced by Messrs. Richardson Brothers, of Hartlepool; but, owing to commercial difficulties, before completion it was transferred to Messrs. John Pile and Co. For this vessel I secured a profitable freight from Havre to Cronstadt. She was to carry three state railway carriages, belonging to the Emperor of Russia, besides 1000 tons of iron shipped at Cardiff. The freight was £2500. She was to bring back an equally profitable cargo, so that in one month this vessel would clear a net profit of £3000. I had gone to West Hartlepool, where I stayed a fortnight to see her finished and hasten her departure. It was on the occasion of this visit that I bought of Messrs. Thomas Richardson, of Hartlepool, and Messrs. Richardson, Duke, and Co., of Stockton, another large steamer. This vessel, of about 1000 tons burden, I called the Marco Bozzaris. It was purchased on the same conditions as those I had previously bought. Mr. George Lascaridi endorsed the bills in my office with the signature of his firm, on the same terms as before.

The Marco Bozzaris soon came to London, and took the berth for Constantinople.

It was about this time that I began to import, for my private account, grain in sailing packets from the Danube. These transactions left large profits. I sent the Marco Bozzaris to Galatz, to take a cargo of Indian corn for my account, which cargo was already sold in advance with a profit of £2000. I had bought the Marco Bozzaris a dead bargain, because of the dulness of trade. Having insured the Admiral Miaoulis for £10,000 more than I had paid for her, to make up her real value in case of a loss —because, as I have already said, I bought this steamer at considerably less than her real value— and having also insured her home freight of £2500, I started for Paris to receive the outward freight on the imperial carriages. I had been about eight days in Paris, and had received the expected money, when a telegram arrived at the Louvre Hotel. It was from Mr. Henry Stokar, my supercargo on board the Admiral Miaoulis. He informed me that the steamer, in a thick fog, had struck on a rock off the island of Osel, and was completely wrecked, but all hands were saved. This vessel had been one month in my possession, and now her insurances and freights, with the stores on shore, left me a profit of £15,000. Yet I was grieved and disappointed at losing her, because she was of a magnificent build and of great carrying capacity, and, in my trade, would have cleared a yearly net profit of at least £8000.

Having returned to London, I received, on account of the insurances, £32,800, independently of the out freight, which I had been paid at Paris. As the acceptances of the Greek and Oriental Steam Navigation Company, which I had given for the Admiral Miaoulis, were to extend over a period of two years, I was able, with the money received for the insurances, to increase the credit of the company and buy fresh steamers. I immediately purchased the Admiral Kanaris, one of the chartered steamers, giving a portion in ready money, and taking credit for twelve months for the surplus. I bought the Asia, of 1093 tons gross measurement, half cash, half credit of twelve months. To these I added the Scotia, 1200 tons; the Modern Greece, of 1000 tons burden; the Patras, of 400 tons burden; and the Smyrna, of 500 tons. For what I did not pay ready cash, I gave the bills of the Greek and Oriental. I contracted with Messrs. Richardson, Duke, and Co., of Stockton, for one large steamer, the Palikara, and three small ones, the George Olympius, the Zaïmis, and the Londos. I contracted with Messrs. John Pile and Co. for the Petro Beys, and with Messrs. Leslie, of Newcastle, for the Mavrocordatos, Leonidas, Colocotronis, Rigas Ferreos, and two small yachts. I bought the Powerful from Messrs. Robinson and Fleming, and the Lord Byron from Messrs. Thomas Wingate and Co., of Glasgow.

It will be seen, by the number of steamers employed, that the trade of the Greek and Oriental Steam Navigation Company had attained considerable expansion. And yet, spite of the number of steamers, I had offers of more goods than I had tonnage to carry them. Notwithstanding the competition of the Liverpool line, I had goods from the best Greek houses of Manchester, besides enjoying nearly a monopoly of the goods of the Greek houses of London. I had even succeeded, some time after, in making the following advantageous contract with the principal shippers of London, by which they agreed to ship exclusively with me:—

[*Translation.*]

THE GREEK AND ORIENTAL STEAM NAVIGATION COMPANY.

To my Fellow-Countrymen and the Shippers of the Greek and Oriental Steam Navigation Company.

At the present moment, when the competition between steamship owners has reached a point beyond which it cannot go, and when our fleet is made complete by the addition of the steamers Smyrna and Patras, which enables us to transport your goods from Holland and Belgium, and place them direct on board our large steamers, without the expense of a double discharging and loading by lighters, I come with more confidence than ever to crave your support. I am persuaded that the few explanatory remarks I am about to lay before you will cause you to continue to use the line as you have hitherto done in so patriotic and praiseworthy a manner.

If this line has done no other good, it has sensibly diminished the monopoly of the freights on your goods, which, up to its formation, was in the hands of a few, and consequently very high. But now that these freights are reduced to their minimum, our competitors, hearing that the two steamers above named are to be added to the line, and finding that you, the shippers, are deserting them, propose, as a catch, to accept such freights as will bring ruinous loss on us.

Leaving to the judgment of each of you the probable results of this step, and encouraged by the promises of the majority of the largest shippers, I propose to regulate the freights according to the following scale:—

Freights.

	Per Ton.	Per Cent.
Sugar (dust) from Amsterdam to London and Constantinople	50/. and 10	primage.
Sugar (loaf) do. do. .	60/. and 10	,,
Sugar (dust) from Rotterdam to London and Constantinople	47/. and 10	,,
Sugar (loaf) do. do. .	55/. and 10	,,

	Per Ton.	Per Cent.
Copper from London to Constantinople . .	30/. and 10	primage.
Iron do. do. . .	30/. and 10	,,
Iron 35 feet in length	40/. and 10	,,
Tin Plates	30/. and 10	,,
Measurement Goods . per ton of 40 ft.	{ 32/6 35/. } and 10	,,
Alum (in barrels)	35/. and 10	,,
Coffee (in bags)	35/. and 10	,,

All other goods in proportion.

If you consent to subscribe to these profitable freights, which are much lower than you used to pay formerly to sailing vessels, I promise—

1. To always have two steamers on the berth, so that no one's goods shall be shut out. The steamers will sail for their destinations every ten days.

2. If any breakage of barrels or other damage should occur through bad stowage, or through the negligence of my superintendent of loading, I undertake to indemnify the merchant. I purpose to get a certificate of good stowage from a surveyor before each steamer leaves the port of London.

3. To give free passage to one or two persons whom the subscribers may wish to send from London to Constantinople, or from Constantinople to London or other ports.

4. To ship free of charge all small parcels or boxes (jewels excepted) which the subscribers may wish to send from one port to another.

With the freights regulated in this manner, and the Greek and Oriental Steam Navigation Company receiving from you the same support that the Greeks of Manchester give to our fellow-countrymen of Liverpool, I shall be enabled to despatch the boats with regularity, and thus prevent all dissatisfaction. You will bear in mind that the company is a national one, and can only continue to exist through your support. This support I now crave more than ever, and I flatter myself that I shall obtain it.

I have the honour to be, Gentlemen,

. Your obedient servant,

(Signed) STEFANOS XENOS.

We, the undersigned, undertake to ship our goods with you at the freights above named.

London, May 18th, 1859.

RODOCANACHI, SONS, & Co.
A. PETROCOCHINO.
G. S. CAVAFY & Co.
VALLIANO BROTHERS.
ZIZINIA BROTHERS.
A. RALLI & Co.
MELAS BROTHERS.
XENOPHON BALLI.
ZIFFO, SON, & Co.
LASCARIDI & Co.
C. & M. SEVASTOPOULO.
THEODORE RALLI & Co.
COUPA, BROTHERS, & Co.
PANA, CREMIDI, & Co.
NICOLOPOULO & CHRISSOVELONI.
A. PSICHARI (Turkish Consul).
CONSTANTINE GERALOPOULO.
SCHILITZI, VOUROS, & Co.

CASSAVETTI, BROTHERS, & Co.
POUTOU & Co.
M. F. MAVROGORDATO & Co.
LEONIDAS BALTAZZI.
H. GEROUSI.
GEORGE G. CEFFALA.
A. FACHIRI & SONS.
J. G. HOMÈRE.
SPARTALI & Co.
ZARIFI, BROTHERS, & Co.
EDWARD VITALIS.
S. CARALAMBI.
T. NEGROPONTI.
M. CASTELLI.
PETROCOCHINO & Co.
D. GEORGIADES.
DELASSO & ALPACK.

CHAPTER IV.

THE FIRST CLOUDS.

At this time Mr. George Lascaridi seldom came to the office of the Greek and Oriental Steam Navigation Company. Months intervened between his visits. The steamers were making money; the grain was leaving a handsome profit. I was really uncomfortable, considering the great help he had given me, to see that he absented himself so much from the office. I one day said something like this to him, and added that the Greek and Oriental Steam Navigation Company was in good condition, and that, if he wished, I was ready to sign a deed of partnership with the firm of Messrs. Lascaridi and Co. He seemed pleased with the proposition. We sent for our solicitor, Mr. Hollams, of the firm of Thomas and Hollams.

I was never more surprised in my life than whilst listening to George Lascaridi talking to Mr. Hollams. He told that gentleman he had decided upon making the Greek and Oriental a limited company, of which he would be a director and shareholder, and he was revolving in his mind other names, thinking who else he would make a director.

I remained silent, and, I must confess, with no slight effort restrained my temper. I did not like to give an Englishman occasion to talk about the " tug of war," and " Greek meeting Greek." When Mr.

Hollams left, which he did without any conclusion being arrived at, I said to Lascaridi:

"Had I foreseen that you would have spoken in that way to Mr. Hollams, I would not have sent for him. Do you think I would be so mad as to turn this company into a limited one? I asked, as before, for a partnership between this line and Messrs. Lascaridi and Co."

"That cannot be," he replied abruptly. "If you like, I'll join you, but not the firm."

"That would be no use," I answered; "you're not a large capitalist. Besides, Messrs. Lascaridi and Co. endorsed the bills I gave to the builders of the steamers. They are the responsible parties."

"They know nothing at all about the bills. I have not passed them in their books," was the cool reply.

"Well, George," I answered, "I must tell you frankly that, spite of the immense debt of gratitude I owe you—for I candidly acknowledge that, but for your generous help, the Greek and Oriental Company would not be in its present splendid position—still, spite of all that, I should feel strong scruples about entering into partnership with you. I do not think I should be justified in doing so, considering that all my transactions hitherto have been with Messrs. Lascaridi and Co. Now, George, this is my position: either I am in partnership with Messrs. Lascaridi and Co., or I am in partnership with nobody. Except all the partners of Lascaridi and Co. give me a letter signed, that they have no claims on the Greek and Oriental, only then will I sign a partnership deed with you. You know what you will have to say to these gentlemen when they come to London and learn all about these transactions, but I really do not know what to say to them."

I have already mentioned that Messrs. Fachri, Lascaridi, and Co., of Constantinople, held coal of mine to the value of £4500; that there had been a loss of some thousands on the chartered steamers; that the Liverpool house of Lascaridi and Co. had charged me a stiff commission on the purchase of the General Williams, though I had bought the vessel in London, direct from the Trustees; therefore, if I now accepted Mr. George Lascaridi as my partner, instead of Lascaridi and Co., what would be my position? In the first place, the entire loss on the chartered vessels would be thrown on the Greek and Oriental Steam Navigation Company exclusively, and Messrs. Lascaridi and Co. would appear as creditors against the company. Secondly, I could not then hold Messrs. Lascaridi and Co., of London, responsible for the accounts of their houses at Liverpool and Constantinople. And, lastly, ties of old friendship with Constantine Lascaridi, brother of George, forbade that I should assent to the proposal made by the latter. In short, I could not, either in justice to myself or in justice to the firm of Messrs. Lascaridi and Co., enter into partnership with George Lascaridi individually.

After the interview with Mr. Hollams, and the conversation that ensued between me and Mr. George Lascaridi, the latter, seeing my determination with regard to the partnership, left the office of the Greek and Oriental Steam Navigation Company, which he never entered again.

Not many days later, I learned with astonishment, from one of the builders of our steamers, that he could make no use of our acceptances drawn by Messrs. Lascaridi and Co., because that firm had accepted nearly a quarter of a million to start the

Galway steam line; that the steamers of that line were worthless as security; that consequently the credit of Messrs. Lascaridi and Co. was ruined in the eyes of the commercial world. He added that, as the Greek and Oriental was believed to be a concern of Lascaridi and Co., the credit of that company was involved in their ruin.

I was thunderstruck by this piece of intelligence. I had not previously been aware of the existence of the Galway line. I put on my hat, and went out to gather what information I could about this company. In the first place, I went direct to the office of Lascaridi and Co. Mr. George Lascaridi was not there. I found waiting for him a Mr. Henry John Barker—one of the Barkers of Smyrna. This gentleman, who afterwards played a great *rôle* in the office of Overend, Gurney, and Co., had been unlucky in his commercial transactions, and was not, at the time to which I now refer, in a very flourishing position. I had met him several times before at Lascaridi's office, and was under the impression that he was looking for some trifling commission business.

As I was very familiar with all the clerks, I asked one privately what brought Mr. Barker so often to the office. He said "he did not know, but thought Mr. Barker had some large transactions with the governor in bills and steamers. He had often seen him there with Mr. Lever."

" Who is Mr. Lever ? " I asked.

" Mr. Lever is the promoter of the Galway line. The Governor has something to do with it."

I saw the clerk knew no more. What I had heard was sufficient to alarm me. If Lascaridi, who knew nothing of maritime property, more especially of mail packets, should abandon his legitimate Mediterranean commerce to enter into such transactions, I saw that he was plunging into the abyss of Tartarus.

On a slip of paper I wrote an invitation, asking Mr. George Lascaridi to dine with me that evening, adding that after dinner we could talk about business. I then took my way through the City, to see what more I could learn about the Galway line.

I was living at that time at No. 27, Victoria Street Flats. Lascaridi was punctual, and after dinner, when we were left alone, I asked him, point-blank, if it were true that he had given the acceptances of his firm to Lever to buy steamers for the Galway line.

"Oh, yes!" he said; "I formed that line also. 'Tis much better than yours!"

"George!" I exclaimed, "in the name of Heaven, do you know what you are about? Do you know that the old steamers on which you have advanced money are worthless? They say that you have advanced three times what they cost. I have heard that Lever paid £8500 to Messrs. Bayley and Ridley for the Indian Empire, and that he mortgaged her to you, a few days after, for £25,000. Is this true?"

"What do you know about all this, and what does the public understand of this scheme?" said Lascaridi, with an air of annoyance.

"Well, Lascaridi, if half the story be true, *addio mastello con tutte di piatto* to the house of Lascaridi and Co. And in what position shall I be and the Greek and Oriental? You know that several of our acceptances, endorsed by your house, are to be renewed by the builders of the steamers. Do you think they will renew if Lascaridi and Co. are obliged to stop payment? Certainly not. And I shall be called upon to pay, perhaps, £50,000 at one moment's notice.

"What have you lost? You lose nothing. You hold the property; mortgage it, and pay," replied Lascaridi, with a sarcastic smile.

"Yes, yes, I know all that; but, once the steamers are mortgaged, good-bye Greek and Oriental. Ah, Lascaridi! you have committed suicide, and you have ruined your partners."

I certainly expressed myself with considerable bitterness against my friend that evening. But Lascaridi was confident that the Galway line would come right ultimately. He even insinuated that he expected to be knighted for having assisted in its promotion.

I tried to get from Lascaridi's own lips some details respecting this blessed Galway Company, that was destined to deal so fatal a blow to many a worthy commercial City house. But he was secret with me, and with the Greeks generally, as the Trophonius; though amongst the English he was in the habit of boasting that he was the promoter of the line.

CHAPTER V.

THE GALWAY STEAM COMPANY.

Messrs. Lascaridi and Co., of London, had drawn several bills on my firm, the Greek and Oriental Steam Navigation Company. These bills I had given to the builders or sellers of my steamers. It was agreed that a commission should be paid to the London firm of Lascaridi and Co. for that paper.

George Lascaridi gave the acceptances of his firm to Mr. John Orrell Lever, who was engaged in establishing the much-talked-of Galway line, which was to annihilate the space that lies between the Old and New Worlds. The scheme was admirable and eminently practicable, and, with the support afforded by the Governments interested, ought to have been crowned with success. But we know how different was the result: the Galway project failed. It was Mr. George Lascaridi who had undertaken to carry out the idea of the promoter; it was he who had advanced, by bills of Messrs. Lascaridi and Co., money to purchase or charter some of the steamers,—he, of course, taking mortgages on them. It was he who had enabled these vessels to start, and show the world the possibility of establishing the line. Ignorant of the nature of steam property, not having the slightest knowledge of maritime affairs or engineering, nor any experience in shipping business, he made the most lamentable mistakes, and he committed unpardonable errors. It is well known that none of the vessels procured for the Galway line were high-priced. What did the Indian Empire,* the Prince Albert, the Pacific, the Antelope, the Propeller, the Adelaide, the Argo, or the Circassian cost?

After all, George Lascaridi might have got over the consequences of making over-advances on steamers had he known how to be silent in season. But no; he opened his mouth when he ought to have sealed his lips, and through his restless vanity betrayed his interests.

The Galway line, like all great undertakings that have had to contend with the prejudices of the multitude, or the private interests of powerful individuals, was at first ridiculed as chimerical. It was so represented by the Liverpool people. The subsidy granted by Government was, in the House of Commons, made a ground of complaint by the Opposition. And yet, spite of all this, the Galway line struggled into commercial existence; it began to attract public notice. It was owing to this line that the steamers of the other companies of Liverpool first began to call at Ireland. Opinions were put forward as to its probable success, and, finally, praise was given to the getters-up of the scheme. This was more than George Lascaridi could bear. He whispered, in hints and innuendoes, that he was the creator—in fact, the capitalist that had advanced money for the undertaking. This becoming generally known, Lascaridi and Co. began to lose credit, and the Galway line *prestige*. How could a line supported by a single individual compete with

---

* Messrs. Bayley and Ridley sold the Indian Empire to Mr. John Orrell Lever for £8500. She was a wooden paddle American steamer, called Henser, belonging to the port of Bremen.

the powerful opposition of Liverpool? Moreover, it was whispered that the capitalist had advanced and taken mortgages on the steamers for twice their value, when a steady man of business would only have taken them for half. And then misfortunes came to complete what want of foresight had begun. The steamers were as unlucky as though the Fate Atropos had secured a permanent berth on board of each. One broke down, another took fire a few miles off the Irish coast, a third ran short of coal, and was obliged to convert her bulwarks, water-barrels, and the covers of her hatches into fuel to enable her to reach her destination. All this was, of course, great fun for the Cunard people; and yet the Galway line was a splendid scheme. Had it been commenced with real capital, and with proper packets—had it been judiciously organized by competent persons—that line ought to have absorbed several of the competing companies.

I have never been able to learn, either from Mr. George Lascaridi or from Mr. John Orrell Lever, the exact amount of the bills accepted by Messrs. Lascaridi and Co. to start the Galway line. One night, as I was reading the manuscript of the present work to Mr. Lever, at his house in Eccleston Square, he corrected me, *en passant*, touching the rumoured £200,000, which he reduced to £112,000. For this amount he was able to account.

Be the amount of these responsibilities what it may—be also the minor details, the petty ambitions, the mutual jealousies, and even the romantic episodes inseparable from the first risings of such an enterprise—be they, I repeat, what they may, one fact is certain, that, as the time drew near when the bills would reach maturity, George Lascaridi became anxious and preoccupied. This was the time that a joint account existed between me and his firm for the chartered steamers. I can now understand his long absence from the office of the Greek and Oriental Steam Navigation Company; I can now solve what was then an enigma to me—his absence of mind, his unconnected expressions, and his remarkable neglect even of his personal appearance.

The Galway line, which figures in the books of Messrs. Overend, Gurney, and Co., as entailing a loss of £839,000, may more correctly be set down at a loss of £1,300,000, because the East India and London Shipping Company, Limited, which is debited with a loss of £578,218, was the legitimate child of the Galway line.* With regard to this latter company, one fact at least is certain, that at the time of its

---

* The steamers whose names follow formed the fleet of the General Screw Steamship Company:—Indiana, Golden Fleece, Lady Jocelyn, Queen of the South, Jason, Hydaspis, and Calcutta. When that line became a losing concern, and was consequently wound up, the European and American Steam Navigation took them up, but also went to grief. Mr. Edward Watkin Edwards, as the official assignee, advertised the steamers for sale, and Mr. John Orrell Lever, of the Galway Company, bought them, through Mr. Edwards; that is to say, Messrs. Overend, Gurney, and Co. discounted the bills of Mr. Howard, of Manchester, and Mr. Lever, and took as security mortgages on these steamers. Mr. Lever, *on dit*, paid about £220,000 for the entire fleet, and then started the Anglo-Luso Company, to run between England and Brazil. I have been unable to learn who it was that caused Messrs. Overend, Gurney, and Co. to be saddled with these steamers, which ruined Mr. Howard and lost no end of money. Shortly afterwards Messrs. Overend, Gurney, and Co., under the advice of Mr. Edwards, established with them the East India and London Steamship Company, which now figures in the estate of Messrs. Overend, Gurney, and Co. for £578,218.

spargana it was ephemerally under the triumvirate of Messrs. John Orrell Lever, George Lascaridi, and Henry John Barker.

Did these three international *hypates* combine, or did two only act in transferring the securities of the steamers of the Galway line to Messrs. Overend, Gurney, and Co.? Did Mr. George Lascaridi alone conclude the transaction, or was Mr. Henry John Barker the sole *proxenites?* I know not. All I do know is, that Messrs. Overend, Gurney, and Co. took up the mortgages on these steamers for the same amount, or a little more, that Lascaridi and Co. had advanced, and for doing this Messrs. Overend, Gurney, and Co. discounted new acceptances on Messrs. Lascaridi and Co., and handed Mr. George Lascaridi a big cheque to meet the old ones. These acceptances were renewed from time to time afterwards.

What advantage did the house of Lascaridi and Co. gain by this transaction? In reality, no substantial advantage; but this financing offered an apparent benefit on two points. In the first place, the immediate ruin of the firm was averted; and, secondly, a unification of its debts was effected, one creditor being substituted for many. The house of Lascaridi and Co. was now in the hands of the great money-lenders of the British empire, and the great money-lenders were under the power of an official assignee of the Bankruptcy Court.

I cannot say who drew the new bills, nor what was the amount, nor can I say for what consideration Messrs. Overend, Gurney, and Co. did the transaction. What I can say is, that once in their hands, the affairs of the Galway Company presented from day to day a new appearance, following the labyrinthian ways and multitudinous intrigues of the wonderful Lombard Street house, whose intricacies will be best understood when I introduce my readers within its portals.

It is not the object of this work to show how long the Galway Company existed under this system of financing, by renewing bills, before it was transformed into the Atlantic Royal Mail Steam-Packet Company; neither is it within the scope of this work to enter into the details of the dispute between the Menclaus, Lascaridi, and the Paris, Lever, which caused that famous arbitration case of many months, decided ultimately by the modern Agamemnon, Edward Watkin Edwards. The prize in dispute, if my memory is not also "a blank," was not a beautiful Helen, but an award of the beautiful sum of £70,000,* which I have never learned if Lever paid and if George Lascaradi received. I do not mean to enter into the petty contests that occurred at the different board meetings of the Galway Company, between Mr. Roebuck, M.P., and my friend George Lascaridi; nor do I intend to notice the negotiations and manœuvres got up to obtain a large subsidy from Lord Derby, and for which our leviathan capitalists killed the goose before she began to lay her golden eggs; nor do I propose to quote the obscure magazine that manufactured sensational paragraphs, eulogistic of the gallant and veteran steamers of the company, and which, strange to say, found their way into the daily press, and kept the Galway line and its promoters in a strong light before the public. Disregarding such topics, I take up what directly concerns the general public and myself—namely, how the great loss of £839,000 must have occurred.

---

* This award some say was considerably more than £70,000.

The Galway Company passed into the hands of Messrs. Edward Watkin Edwards and David Ward Chapman, "perfectly solvent," as Mr. Lever asserted in his letter to the *Times* of the 28th of January, this year.

Mr. Edward Watkin Edwards, though he had never been a seaman, a merchant, or a naval architect, undertook to organize the company. He proposed, with, of course, the consent of the new company's board, to build five large steamers, having all the modern improvements. With these he hoped to retrieve past losses, to fling a halo of fame round the name of the company, recover public confidence, and benefit the capitalists and shareholders. The Atlantic Royal Mail Steam-Packet Company constructed the Hibernia, the Columbia, the Anglia, the Connaught, and bought the Adriatic. If I do not mistake, these steamers cost from £80,000 to £120,000 each;* but, instead of

---

* The new steamers bought by the Atlantic Royal Mail Steam-Packet Company were the Hibernia, 2005 tons, 800 horse-power, built on the Tyne by Palmer Brothers, at a cost of about £100,000. Messrs. Laird, of Birkenhead, were afterwards paid £40,000—some say £60,000—to strengthen her. If such be really her cost, she would not realize to-day £20,000. Here is a dead loss of £140,000 in one single steamer, exclusive of what she lost on her voyages. The Columbia, 2903 tons register, 1000 horse-power, was built in 1861, at Hull, by Messrs. Samuelson and Co., as well as the Anglia, 2949 tons register, and 1000 horse-power. These steamers cost, as I am informed by a competent authority, about £100,000 each. The Adriatic, 3670 tons, and 140 horse-power, was an old wooden American steamer, built by Mr. George Steer, of New York, in 1854. The Connaught, which was lost, was built by Messrs. Palmer Brothers. These steamers were mortgaged to the Imperial Mercantile Discount Company, Limited, of which Mr. Henry John Barker, formerly of 4, Abchurch Lane, was the manager. All these steamers, in which such an enormous amount of money was sunk

profits, each voyage entailed losses. Had these vessels made profitable voyages, they would be still running. The consumption of fuel by one of these steamers is something fabulous. I saw three of them lately lying at Liverpool; they are fine-looking vessels, and were they serviceable, the Liverpool people would not allow them to lie idle. I doubt whether a purchaser could now be found that would give £20,000 a-piece for them.

Now, five such steamers would cost £600,000. And whence came such a sum to be locked up in so depreciating a property? Whence came the working capital—no trifle—of this line? What became of the old ships, and how much did their mortgages realize? Who bore the loss they entailed? What moneys had the Atlantic Royal Mail Steam-Packet Company from the public?—what sum did the shares realize? If all these items are put together, it will not be difficult to divine what became of the £839,000. In such a colossal business there are always small commissions, brokerages, douceurs, charges for bonus, interest, &c., which tend to draw off the vitality of the concern.

---

are, with the one exception of the Connaught, now eating their heads in the docks of Birkenhead and Southampton, and are quite unmarketable. I have heard that, within the last few days, the Anglia has been sold to the Turkish Government for a great deal less than £20,000.

CHAPTER VI.

A COMMERCIAL PERSEUS.

Mr. GEORGE LASCARIDI, spite of all his commercial blunders, is a most noble-hearted and upright man. His generosity is such that, had he the means, he would willingly, at his own cost, relieve all the wants of suffering humanity. He is extremely ambitious, but his ambition is blended with a soft-heartedness that renders him unable to say " No," even to those whom he knows are imposing on his kindness. At the head of a powerful Greek firm, he grew tired of following the monotonous routine of the export and import Greek trade; he wished to distinguish himself in the British metropolis by some great national act that would win the praise of the English people. Had the Galway line worked successfully, Lascaridi's foresight, pluck, and enterprising spirit would have been the theme of universal praise. Besides the wealth he would have reaped, so splendid a success would have paved the way for his firm to enter on still greater undertakings. It is sad to think the result was so different.

Commercial men, even of great experience, often commit serious errors. Whether George Lascaridi was misled by others, or whether he voluntarily took a leap in the dark, I cannot say; but I am convinced that what he did was done in good faith, and in expectation of surprising his partners and fellow-countrymen by his splendid success. He could not suspect others of malicious practices of which he was not himself capable. He was ignorant of the maxim that now obtains amongst the City neophytes: Ὁ γὰρ θάνατός σου ζωή μου (your death is my life). Entangled in a many-meshed net, and seeing that his partners were irritated because he had involved them in affairs from which they had wished to keep entirely apart, his sole aim was to get out of the difficulty as best he could. Was it pardonable in the promoters of the Galway Company to take advantage of his ignorance of maritime property, and saddle him with worthless steamers? If so, where was the harm in his transferring the precious load to the shoulders of the first capitalist he could find ready to assume the burden? Unfortunately, my friend Lascaridi, carried away by a strong ambition, not satisfied with engaging in the Galway Company, embarked simultaneously in other enterprises, and became a veritable commercial Perseus. If he never killed a Medusa or rescued an Andromeda, it is certain that he petrified more than one Polydectes. Simultaneously with the Galway affair, he got into commercial connexion with Mr. Valsamis, a Greek merchant of Glasgow, and his relative. To this gentleman he also gave the acceptances of his firm for some thousands of pounds, wishing to extricate him from certain difficulties. He was equally liberal of these acceptances to some Greek firms of Manchester. It seemed as though he would fain extend his arms to every corner of the earth. He sent Mr. A. B. Manuel, one of his clerks and a distant relative, to Gibraltar, and expended thousands in bale goods to clothe the Moroccians and Tunisians. The acceptances were also brought into requisition to assist a M. Mavrogordato to establish, in different streets of London, French glove-shops, as well as French perfumeries. Of all these speculations, the most useful was the famous bill-discounting office of Mr. Henry John Barker, at No. 4, Abchurch Lane.

Lascaridi was precipitate in action. He often, it is true, made a lucky *coup* that carried him to the very colophon of success; but if the blow missed—if his undertaking were attended by ill-luck—then his reasoning faculties seemed to be obscured. He was sometimes fickle as the wind; at others, obstinate as a mule. A flatterer could turn him away from the most serious business, when a friend could not open his eyes to a flagrant impending calamity. He was mysterious about trifles, and foolishly communicative in cases where his interest demanded profound silence. He would trifle with his commercial credit as a virtuous but vain woman often does with her reputation when she flirts with a snob. He associated himself with the most chimerical schemes, and endeavoured to carry out at the same time ten different projects. What says the Greek proverb: "He who pursues many hares at once, catches none." And so it was with Lascaridi.

Mysterious about trifles—flippant when dealing with serious business—most cautious that the world should not know that he paid four shillings for his dinner—he would, at the same time, not hesitate to boast publicly that he had advanced £200,000 to promote the Galway Steam Company, or that he had lent £100,000 to establish the Greek and Oriental Steam Navigation Company. What I think will astonish the world is, to hear that, in advancing all these acceptances of Messrs. Lascaridi and Co., in engaging in these large undertakings, Mr. George Lascaridi seldom made a written agreement, seldom had a distinct verbal understanding before witnesses, and seldom had a moral guarantee or security from the persons with whom he thus connected himself. He launched into wild speculations, he plunged into fathomless commercial depths, but he always hoped to recover his footing and come safe to land. He did not like written contracts, because the men whom he put forward or assisted were poor men, peaceful and honest, but his creatures. As they were entirely dependent on him, he thought that any time it was in his power, if things turned out well, to take his share of the profits; because, in the always possible event of bankruptcy, it was obviously more advantageous for his firm to appear as a creditor than as a partner.

Did George Lascaridi seek to imitate Marc Antony, who, when his indebtedness amounted to £500,000, cleared all in one swoop, leaving Rome to wonder where he found so much gold? Or did his ambition make him aspire to resemble Marcus Crassus, who, after feasting the Roman people, and squandering money right and left, died worth a million and a half sterling? And was it in the spirit of the famous Claudius, who was wont to give fabulous prices for slaves whom he designed to set free, that he took by the hand Mr. Henry John Barker, M. Valsamis, M. Manuel, and others, intending to make them wealthy as Pallas, the favourite freedman of Claudius? Mr. George Las caridi, in my humble opinion, was flying on a Pegasus like a commercial Perseus.

CHAPTER VII.

MY AGENTS.

THE Greek and Oriental Steam Navigation Company was now entirely under my control. In fact, that company was now nothing more nor less than myself: I was sole owner, sole manager, and sole director. And what times those were in which that company came out—what prospects were opened before it! I avoided competition in London by opening the line of the Danube, and, instead of sending my steamers on to Odessa, made them touch at the intermediate ports; also by making my contract with the Greeks of London. With respect to the home cargo I had nothing to fear. The Greek shippers in the Levant not only regarded my line with a patriotic affection, but even went so far as on some occasions to pay five shillings per ton higher freight to my steamers than my opponents were asking. But this was not all.

I had sent up the Danube several small steamers, each of about 2000 quarters of grain-carrying capacity, and drawing a small draught of water. It was intended that these small steamers should go up the river as far as they could float, load the grain there, and bring it down to the Sulina mouth, where my large steamers, having unloaded the cargoes they brought from London, would re-ship the grain, and bring it back to England. I considered that, if other Greek houses were making large profits, though they bought the grain at Galatz or Ibraila, and shipped it on chartered vessels, I ought to be able to do much better, seeing that I could buy the grain many miles higher up the river, and directly from the growers. Besides, being able to transport the grain by steam from Sulina to England, I gained in time on my opponents.

My plan was comprehensive, and success inevitable. Convinced of this, I did not delay putting it into execution. I immediately sent a Mr. Henry Stokar to Galatz to take the management of the Danube steamers, and buy the grain. I furnished him with large credits.

Mr. Stokar was a young man who had been some time in my employment. He was supercargo on board the Britannia, then clerk in the office, then supercargo on board the Admiral Miaoulis; and when that vessel was wrecked in the Baltic, he behaved so gallantly that I gave him the appointment at Galatz in preference to many who perhaps thought they had stronger claims. He was of mixed race—his father being French, his mother English—and possessed of great courage. The Gallic spirit revealed itself strongly in his nature. He was a proud, high-minded man, hot tempered, and apt to regard things from a military rather than a commercial point of view. He often construed a slight pleasantry into an offence, and was always ready to demand gentlemanly satisfaction. He had never been engaged in business, and had no commercial experience, but he was a shrewd, sagacious man, very persevering, and most zealous in the interests of his employer; and, over and above all this, he was a strictly honest man. But, though he spared no trouble, and even risked his life in my service, he unfortunately stuck to his orders in so strictly military a spirit, that he would never deviate therefrom, even when circumstances had arisen that made the fulfilling of them a palpable loss, or the departing from them an evident gain. Another of poor Stokar's weaknesses was his economy. This he carried to such an extent at Galatz, that he would fain do all the business himself. He never engaged a proper staff of clerks; he had not even a book-keeper.

After Stokar had been some time at Galatz, the effects of his system—or rather his want of system—became apparent. I, who did not know where the error lay, was quite puzzled. Though Stokar had shipped and consigned about thirty thousand quarters of Danubian grain, he had never sent me any account or invoice for what he had paid, so that I was left in utter darkness as to how the machinery was working. His letters were very long, full of projects, hopes, and expressions of devotedness, but they were very uncommercial, self-contradictory, and obscure. I did not know what to think. I sometimes began to fancy that Stokar had fallen into the hands of those men called *magazzinieri* (warehousemen), and that they had deceived and robbed him, or else that he was not the man I had taken him to be. Stokar applied to me for fresh credits before he could ship all the grain he had to England. I became alarmed. Knowing that the best-conceived plans are often rendered abortive by a check suffered in the commencement, I decided upon immediately sending some one to Galatz, to inspect affairs there. I selected for this task a M. Peter Theologos, a Greek Smyrniote, whom I had some months before engaged for the Greek correspondence. This M. Theologos had some commercial experience, having been formerly connected with a celebrated firm that bore the name of the great Jewish lawgiver—that firm which a spendthrift nephew abandoned to the direction of clerks, whilst he amused himself, at the West-end or at Paris, in pursuits more congenial to his age and tastes—that firm which fell with a crash that re-echoed through England and the Levant. I may just observe, *en passant*, that when the wardrobe of a certain fair Timandra was brought to the hammer, twenty-five magnificent new dresses, made by Madame Rape, of the Boulevard Italien, were knocked down at £800—an enormous price. After the fall of that house Theologos started in business on his own account, and floated on the ocean of commerce like a bit of cork that never sinks, but that never reaches harbour. He was in great distress. One day he came up to me in the Corn Market, and told me his desperate position, and asked me to lend him £10. It was the 15th of June, 1858. I did it with pleasure,* and I afterwards took him into my office at a salary of £150 per annum.

Theologos was a thoughtful, taciturn man. He smoked little cigarettes of Turkish tobacco whilst he wrote his letters, but his work was well done. He never made any observation, never advanced any opinion; he was of a tranquil and contented nature, and seemed to take the greatest interest in the business. It was about the time that Theologos entered my employment that Mr. Alexander Carnegie, finding he had not prospered in business, closed his own office, and entered mine, as head clerk, at a salary of £350 per annum.

I had known Alexander Carnegie since 1856. He was a clerk in the office of the ship-brokers, Messrs.

---

* Received of Mr. S. Xenos the sum of ten pounds, which I promise to pay after three months. London, 15th June, 1858.

P. THEOLOGOS.

A. Blackie and Co., 23, Harp Lane, Great Tower Street, with which, since 1853, I have had large transactions in ships and coal. He started in business on his own account as ship-broker, in 1856, at No. 2, Ingram Court, Fenchurch Street. I chartered through him the Youth, Token, Foreman, and some other ships, which turned out most profitable.

Alexander Carnegie was then a young man. He was very active in his movements, and endowed with an extraordinary perseverance and volubility of tongue, which he delighted to display upon any subject of which he was master, or of which he believed his auditory to be ignorant. He was an ambitious and very sharp man.

Carnegie tried to join me in business in the year 1856, seeing my connexion with the Greek shippers. We spent many evenings together. He wanted to arrange with me that all the loading business of my packets should be transacted in his name for my accommodation. I declined. I was, however, always very willing to give him any business upon which he could make a good commission. In 1857 our acquaintance declined. He continued a ship-broker on his own account, and after three years (the 15th of March, 1859) he found himself landed in my office, where fortune waited him. (See Appendix, No. 8.)

When Theologos left for the Danube, I gave him letters to Stokar, authorizing him to inspect the books. I also gave him private letters to the captains, recognizing him as my agent, and I gave him private letters for Stokar, ordering him to deliver up the business to Theologos and return to London. These private letters were to be used only in case he should discover some abuse of confidence or breach of trust on Stokar's part.

Theologos's letters in the beginning represented my Danubian affairs as in a state of great confusion. He said it was difficult to put matters in order, on account of the absence of books and vouchers. His reports gradually assumed a darker colouring, until finally he delivered his letters of recognition to the captains and to Stokar. The latter had, from the first moment of Theologos's arrival, complained in his letters to me. He had evidently taken a dislike to the man, whom he regarded as a rival. It was no wonder. They were totally opposite characters. One was open, impetuous, untouched by the spirit of diplomacy; the other was silent, deep, self-possessed, mild-looking, and endowed with great diplomatic tact. How often does it happen that, in obscure and narrow circles, as subtle a game is played, and with as much skill as was ever displayed in the most Machiavelian cabinets. In such a struggle, the victory is to the calm. Poor Stokar was doomed to fall.

When Theologos took the command of the business, Stokar's rage was so great that he refused to deliver the sums of money that he held. Nor was it until the interference of the British consul was obtained that he gave up the deposits of coal I kept at Galatz. Even then he did not deliver the coal to Theologos; he placed it at the British consulate, and came to London to settle accounts with me in person.

If Stokar had delivered up the business and the property of mine that he held to the man I sent out, and then come to London and entered into explanations with me, there is no doubt but that I should have sent him back to Galatz to manage with Theologos that well-planned scheme which within twelve months revolutionized the grain market of Galatz. But instead of doing this, Stokar went about the little town of Galatz exposing my affairs, and, consequently injuring my credit. He tried to keep as security the property of mine that he held. All this *esclandre* put arms into the hands of my trade competitors in those parts who exported grain, and their consignees in London.

Before Stokar arrived in England, I had sent Carnegie out to Galatz, where I established the firm of Theologos and Carnegie. When Stokar came to London, he tried to give me a verbal explanation of how he had managed affairs at Galatz. But, as a commercial man, I could not be satisfied with verbal explanations. I required arithmetical evidence. Stokar could not produce a satisfactory balance-sheet. Besides, his late proceedings at Galatz had so shaken my confidence, that I could not think of again taking him into my employment until he should have satisfactorily explained his accounts. Stokar became excited, impertinent, and menacing. He threatened law. I was willing to make every excuse for the vexation he felt, and, to avoid litigation, we agreed to submit our differences to an arbitration. Stokar claimed £4000 from the Greek and Oriental. However, I appointed an arbitrator, Stokar named another, and an umpire was agreed on. This arbitration business was prolonged throughout a year, and might have gone on still longer, but that I happened one day to meet Stokar on board a penny steamer on the river. I asked him to dine with me, and that evening we arranged the matters in dispute. I agreed to give him £1000, and each was to pay his own expenses. The arbitrators were taken by surprise. They had sat, like two Vice-Chancellors, disputing points of law, when a little common sense would have settled the business at once. I must say that I never paid money with more satisfaction than I did that £1000.

Poor Stokar was victimized. He had been robbed by the *magazzinieri*, but that knowledge was suppressed, and all the blame was thrown on Stokar.

When Stokar found himself in possession of £1000, he returned to Galatz, and attempted some commercial speculations on his own account (he inherited also, one year after, £3000 from a relative); but after a few years he returned to England penniless. Within a few weeks after I read the announcement of his death in the *Times*. The intelligence sent a pang through my heart. Stokar, with his chivalrous sense of honour, would have been an invaluable servant to me, had I sent a good book-keeper with him to Galatz, and given him my brother-in-law as a partner, to supply his want of commercial experience. Had I done this, I should never have fallen into the hands of Overend, Gurney, and Co., and suffered the annoyance and heavy losses caused me by the ungrateful clerks that took his place.

CHAPTER VIII.

A GRAVE MISTAKE.

I SHALL now return to the working of the great machinery that I had established in London, and which forms the principal topic of this narrative. To accept bills for £150,000 is only an affair of a few minutes; to provide for them is also only a momentary concern, if your credit is virgin, your property free, your business well organized, and your enemies too insignificant to do you serious injury. But it is a Sisyphean labour if, while you are trying to meet your responsibilities by financing, you become the object of a Government opposition. For a short time I was in the first category. I had established my credit, I had overcome my enemies, and I thought all was safe. I hoped, with a fleet of steamers, I should be able to realize numerous national schemes which I had formed. I was, besides, the owner of a Greek newspaper as powerful in the Levant as the *Times* in England, which supported the steamers. What ought not the possessor of such means to be able to achieve with due capacity? But the change of the commercial position of George Lascaridi now, and the conduct of Theologos and Carnegie, caused the failure of my mission. Had I known then what I know now, I would have given a different turn to affairs. I could have saved the Greek and Oriental in 1860, by turning it into a limited company. I could at that time have secured the support of many powerful personages to join it. Had I done this, I should never have fallen into the hands of Messrs. Overend, Gurney, and Co.; I should never have been in the power of Edwards and others, who, after having fattened on the ruins of my business, when no ties of interest united us any more, pretended that they did not even know me.

When the first bills fell due, they were, of course, punctually met, and there still remained a large balance at my bankers; I could, however, see that before three months I should be on the brink of a precipice. Lascaridi and Co.'s credit was gone; I had nothing to expect from them; in fact, I could never now get a glimpse of George. His time was wholly occupied with the Galway Company, which had involved him in endless arbitrations.

To meet my difficulties two ways were open before me: I could either "finance" the company, or raise money by mortgaging the steamers. Money was at that time very cheap, being at about $3\frac{1}{2}$ per cent.

"To finance a company" is a technical phrase in use amongst the great City capitalists, and means nothing more than raising money after a certain fashion. Some person connected with the company that is to be "financed," draws bills, either from abroad or in England, upon the said company; this paper is discounted by some bill-broker, and the money handed over to the company to meet urgent payments. My company could be "financed" much more easily than is daily done in other cases. The foreign exchanges might have been made the basis of the financial operations. Such a mode is often more profitable than drawing accommodation bills, but it requires credit, and that your partners abroad should be shrewd and active, and know how to read the barometer of foreign exchanges.

I did not intend to adopt that practice, because my agents were not of a capacity to carry out such an operation. What I proposed was, that my agents at Galatz should draw at three months on the Greek and Oriental Steam Navigation Company, for the full value of the grain shipped on board my steamers. On the receipt of the bills of lading I could sell the cargoes whilst still floating, and receive the money long before the arrival of the steamers. By acting in this way I should have in hand the value of the grain and the home freight—viz., larger sums of money than my monthly liabilities amounted to.

By this mode of proceeding I should also escape the commission and heavy interest that must be paid to mortgagees and to bill-discounters.

The realization of my scheme depended entirely on the capacity, the tact, and the good commercial management of my agents at Galatz. I could not attempt to enter into arrangements with any of the Greek houses that had an establishment at Galatz, because, in doing so, I should be obliged to give details concerning my business—I should be obliged to tell how I had purchased my fleet, who were the capitalists concerned, and who my partners. In fact, I should have been obliged to solve the enigma that puzzled the whole Greek trade; and what would be the consequence? A revelation of my affairs would have shaken to the foundation the splendid fabric that awed and dazzled so many. Any other Greek house, on learning that I had out acceptances for so many thousand pounds for the value of the steamers, would have told their correspondents to have nothing to do with my paper.

A propos, I forgot to mention an instance in which the inexperience that had thrown me into the clutches of Messrs. Overend, Gurney, and Co., also caused me serious loss.

Before the outbreak of the Italian war, and before my connexion with Messrs. Overend, Gurney, and Co., in December, 1858, I was owner of 50,000 quarters of Black Sea grain, principally wheat, unshipped or shipped in ships and steamers, and 30,000 tons of coal. This grain cost me about 32*s.* per quarter. Suddenly prices went up to 55*s.* per quarter, and coal £1 more than its cost price. Here were the means of realizing £80,000 profit by an instant sale. Calculating upon what had occurred during the Crimean War, when the prices of wheat went up so rapidly to 80*s.* per quarter, and coal to £4 per ton, I expected that within a few weeks prices would be doubled, and that I should realize a sum sufficient to clear all my steamers. My brother Aristides one day called upon me to induce me to sell. "Sell, Stefanos, and you will never repent," he said. "I will not," I replied; adding, "The market in a few weeks will be 10*s.* or 15*s.* per quarter higher. At the time of the Crimean War wheat went up to 90*s.* and this war will last many months." "Sell, Stefanos; if I were in your place I would sell, at the present profit, the corn and the coal." "No," I replied again. I played *tout pour tout*. The suddenly concluded peace of Villa Franca caused a revulsion in the grain and coal markets. Prices rapidly returned to their former figure. So vanished, within a few days, my fairy vision of gain. Instead of £100,000, as I hoped, I only made on the transaction about £10,000. If I had sold, I should never have fallen into the power of Messrs. Overend, Gurney, and Co.

CHAPTER IX.

DISAPPOINTMENT.

THE footing on which I was standing at that time was so much better than that of any other commercial Levantine house, that I ought to have had no difficulty in obtaining the credit I needed. My company was not limited, it was unlimited, and I was the sole transactor of its business—the sole owner in whose name the steamers were registered at the Custom-House, and the sole acceptor of the bills. The steamers were not idle—they were constantly in full work; they were free of mortgage, and represented a large tangible capital. How different was such a position to that of the many houses that just then began to inundate the grain trade—houses that, under the brilliant robe of credit, hid a skeleton form!

My agents, Theologos and Carnegie, who were to be the instruments for the carrying out of my project, had at Galatz several steamers in full work; they had large deposits of coal, representing a vast capital, and they startled their corn merchants by the large credits I had given them on first-class London bankers. With such credits, they ought gradually to have put the paper of my company into circulation, and so have raised my credit to the highest point. I had a clear understanding with them on this subject before they left England. They knew perfectly well that the projected credit was the real steam that was to keep my fleet afloat.

I am not vain-glorious; but no practical merchant can fail to see that my scheme, fully worked out, would have brought in a golden harvest. But it is not sufficient for a merchant that his project is well devised; it is equally important that he should find servants competent to carry it into execution, and conscientiously devoted to their employer's interest.

I was more unlucky in my second choice of agents than I had been in my first. But the worst was, that I had placed myself in a position in which I could make no move now, nor remove my agents, without disturbing the action of the whole machinery. When I wrote to them in Galatz, ordering them to draw on the company for some thousands of pounds—part to buy grain, and part to be remitted to me in drafts on first-class houses, to enable me to meet some urgent payments—I received an absolute refusal from Carnegie, who finished his letter by telling me that he had not gone to Galatz to furnish me with money, but to make some for himself. He further added that I must send him fresh bankers' credit.

My disappointment was great. My first impulse was to recall Carnegie. He was not the man to develop a scheme which would eventually have brought him wealth, and raised him in the social scale. It was plain that he did not wish to row in the same boat with the man who had pulled him half drowned out of the water. I immediately telegraphed to my brother-in-law at Smyrna, asking him to take the management of my house at Galatz. He was then a merchant in easy circumstances, and possessed of large landed property in Smyrna. He was so indignant because, when I removed Stokar, I refused his demand to go there, that now he declined the offer. I was at a loss who to send out with Theologos, when I received a letter from the very man, saying that he wrote, at the desire of his partner, to say how sorry Carnegie was for his last letter. They were both, he said, willing to do whatever I desired, but that the fact was that Stokar had circulated so many evil reports against me and the company, that they could only establish the credit by degrees, which they were very anxious to do. So he proposed to draw two-thirds bills on bankers and one-third on the Greek and Oriental. The same mail brought me a letter from Carnegie, expressing his regret for what his former epistle contained. I had also a letter from Scotland, from Carnegie's father. He asked me not to remove his son, because he was devoted to my interests, and personally attached to myself.

An excess of prudence has ruined many men. Remembering what took place in Stokar's case, I feared that, if I removed Carnegie, he would, perhaps, go open-mouthed through Galatz, and injure the credit of the Greek and Oriental Steam Navigation Company; so I resolved to swallow the pill. On the other hand, I must confess that the grain sent to England by Theologos and Carnegie was all of good quality, and bearing a surplus measure. Then the bills of lading of the cargo were accompanied always by a correct invoice; so that I knew to a penny what the grain cost.

The prospects of Theologos and Carnegie were excellent. I gave them £1000 per annum for the expenses of the office. They had also a commission of $2\frac{1}{4}$ per cent. on all the grain they shipped, and a commission of $2\frac{1}{2}$ per cent. on all the goods imported to Galatz by my steamers. If, therefore, they drew on

London bankers instead of on me, they incurred no risk whatever. Being empowered to draw bills on London nkers, all the bill-brokers, merchants, and bankers of Galatz that wanted to remit money to England would have been suppliants before them, asking to buy paper which they knew to be the best. They would have thus been placed in an influential position—one which would be at the same time perfectly safe and comfortable. But, to obey my instructions in the carrying out of my project, they would have been obliged to work hard, and they would not be in the beginning great personages.

Another point that obliged me to submit to Carnegie's insulting letter was, that they held property of mine to the amount of between £15,000 and £20,000 in grain and coal. They might have raised claims, as Stokar did, and some question about the possession of these goods, and thus involved me in another arbitration case.

I understood my position perfectly well. I saw that I had nothing to depend on but my own exertions. There were, however, two circumstances in my favour. Money was cheap, and the machinery of my scheme was working in London perfectly well, having abundance of cargo.

CHAPTER X.

THE ULTRAMARINE POWDER.

IF, in order to vindicate my own character, I shall find myself compelled to lay bare errors committed by other people, I shall not, at all events, expose myself to the charge of attempting to hide my own faults. The difficulties which compelled me to seek the assistance of money-lenders arose from two causes. Firstly, I had under construction several steamers, and to the builders of these I had advanced the acceptances of the Greek and Oriental Steam Navigation Company. Now, these steamers would not come into my possession until several months after the bills should have reached maturity. Here was one difficulty. Secondly, I was obliged to give security to bankers in order to obtain credits for my house at Galatz, so that my agents might have the means of purchasing grain, that branch of my trade which had originally yielded such large profits.

When I contracted for the construction of the steamers, the house of Lascaridi and Co., with which I was acting for a joint account, was in the zenith of its credit. I was wholly ignorant of the transactions in which George Lascaridi was engaged, nor was I aware of his misunderstanding with his partners. It was an important point to establish the credit of my Galatz house in the Principalities, because, if my agents there could make their first purchases with cash, they would be afterwards able to load our steamers in the same way that the other Greek houses were loading their sailing ships.

In taking these two steps, I had reckoned on assistance from George Lascaridi, the credit of whose firm was then unsullied. When I found myself disappointed, I was exasperated. Wrathful collisions ensued between him and me. When I reminded him of the approaching maturity of some of the bills endorsed by his firm, he referred me to Mr. Henry John Barker, to "finance" me, as he called it. I did not like this new Pylades that my friend had taken to himself, but I was not in those days aware that Lascaridi was the real founder of the little bill-discounting office at No. 4, Abchurch Lane.

I have said that I do not intend to hide my own errors and defects. I here confess to a prejudice—I dislike English Levantines. They do not inherit the Anglo-Saxon firmness of character nor the Greek vivacity of intellect. Brought up in the Turkish school, where advancement is obtained by bowing and scraping, by fawning and favouritism, they become adepts in these arts. They are neither English, Greeks, nor Turks, but a mixture of all; they can change the hue of their nationality with a chameleon-like facility to suit emergencies—a facility that enables them to stand the fire of commercial battles and defeats with salamander-like indifference.

I dislike smooth-faced, soft-voiced men, that have not the moral courage to look you straight in the face and utter a solid "no" or a sterling "yes;" men who are always ready with the pretext of important business, and run, hat in hand, out of their office the moment they see you enter, if they fancy you come to ask a favour; or else allow you to do "long antechamber," as the French have it—robbing you of your time, when they have already determined not to do what you want. And yet, if these men want a favour, they will stick to you with the tenacity of a dog holding a bone, until they serve their ends, and when their object is accomplished, and they no longer want you, they pass by as if unconscious of your existence.

Mr. Henry John Barker introduced me to Mr. Albert Gottheimer, who, with Mr. Josiah Erek, managed the Mercantile Discount Company, Limited, 24 and 25, Birchin Lane.

Mr. Albert Gottheimer, formerly a wine merchant trading under the firm of Coverdale and Gottheimer, was at that time studying closely the LIMITED SYSTEM, through which the blind sister of the Fates afterwards made him a magnificent GRANT. Mr. Gottheimer was willing to discount my promissory note of £3000 for one week for the premium of £100, provided I gave him as collateral security a bill of mortgage on one of my small steamers, which he pledged his word of honour not to register in the Custom-House until the promissory note came to maturity. He further kindly promised that, should it not be convenient to me to take up the *petit billet* at the expiration of the time, he would renew on the same terms, and continue to do so, provided I invested £1000 or £1200 in the shares of the Mercantile Discount Company, Limited, which was paying, he said, 15 per cent. per annum.*

The future father of the Credit Foncier and Mobilier of England was too sharp for me. I took the shares to oblige him, and they soon burned my fingers.

---

* The capital of the Mercantile Discount Company, Limited, was £200,000, in 4000 shares of £50 each (£25 per share to be paid up). Out of the 4000 shares very few were taken up.

The little office in Birchin Lane, anxious possibly to shoe the public comfortably, discounted so many bills for the leather merchants, that when, in 1861, a panic shook the hide trade, down tumbled this *primaroli* limited company. It was there Mr. Albert Gottheimer acquired all the knowledge and skill which, in 1864, Mr. Albert Grant displayed to a delighted metropolis.

When Mr. Albert Gottheimer withdrew from the Birchin Lane concern, he compelled the shareholders, in virtue of the Articles of Association, to pay him a few thousand pounds. My acquaintance with this gentleman was of very short duration. Saddled by him, in the way I have described, with a *blue powder*, called "ultramarine," which I took by paying him £200, I sent it by one of my steamers to Constantinople, to be sold, as he told me it would be, at a great profit. Who would believe that the descendants of Aaron or Moses —the only buyers of such stuff in Turkey—would not offer even £5 for this, though the freight, insurance, and shipping charges amounted to more than £10. Finding the precious article unmarketable at Constantinople, I sent it on to Galatz, where it met with no better success, and where, I believe, it remains to the present day, not worth the expense of being shovelled into the Danube.

CHAPTER XI.

THE FIRST MORTGAGES.

MR. GOTTHEIMER had a head too long for me. I was so disgusted with this mode of "financing," and so tired of Barker and Lascaridi, that I resolved upon putting an end to the state of anxiety in which I was kept. I decided upon calling a meeting of the shipbuilders to whom I was indebted, and the holders of my large bills, and coming to terms with them for the balance due. I intended to promise them 20s. in the pound, with 5 per cent. for long credit, say two or three years, giving some of the steamers as security for the payments. During these two or three years the steamers, working steadily, would gradually free themselves. This arrangement would leave me at least two or three steamers free as a reward for my hard work. I sent for Mr. George Lascaridi, and communicated my determination to him. He was seized with terror. He reminded me that the bills had been endorsed by his firm, and asked if I wanted to break down Lascaridi and Co. and ruin him, as, of course, the builders would look to his firm for payment. He appealed to my sense of generosity. I was between two fires. Had I been born with a foreknowledge of events, I should have known that my proposition was best both for Lascaridi and myself. As it was, I yielded to him. And now he, whom I had not seen for months, set his wits to work to raise money. He proposed that I should mortgage the Admiral

Kanaris and the Marco Bozzaris; he despatched his grand emissary, Mr. Henry Barker, to seek, *per mare et terrâ*, the precious metal of which we stood so much in need. Mr. Barker succeeded in discovering the golden sands of the Pactolus of Lombard Street. With oriental ability he succeeded in throwing a pontoon from Abchurch Lane to the great "Corner House." He did more. He contrived to fix a tube through which he whispered into the ear of the proud David Ward Chapman. Having made these arrangements, he informed Lascaridi that "he could do the job for 20 per cent." I positively refused, preferring my own plan of coming to an understanding with the shipbuilders, who, when they should have learned all, would be happy to accept the security I offered instead of the simple bills they held. Barker and Lascaridi redoubled their exertions. The day after my refusal I received the following note in Greek from Lascaridi :—

DEAR STEFANOS,

Overend, Gurney, and Co. agree to make a loan for six months on the following terms :—£19,000 to draw on the company at six months; £1000, their commission, for six months. They will discount the bills for £19,000 at 5 per cent. interest. I make my calculations that it will be altogether 15 per cent. per annum. I think you must swallow this pill. It is impossible to get the thing done cheaper.

Your friend,
18/2/59   G. P. L.

Lascaridi forgot to add the small commission to Mr. Henry Barker for his trouble, which would make more than 15 per cent.

Bound by strong ties of gratitude to Lascaridi, I accepted these conditions rather than bring about a collision between him and his partners. Within a few months he and Barker concluded with Messrs.

Overend, Gurney, and Co. fresh mortgages on three, small steamers for £15,000. The terms on this transaction were stiffer than on the former—£2000 bonus, and 5 per cent. All this time I had not made the acquaintance of Messrs. Overend, Gurney, and Co. They advanced the money to Barker, and he paid it over to me in instalments. I was soon tired of this zig-zag mode of doing business. I began, too, to see that the Greek and Oriental, far from being saved by this "financing," was running rapidly along the high road to perdition.

What was my surprise when one day the smooth-faced, soft-voiced Mr. Henry Barker asked me to give him a second mortgage on one of the steamers as security. He had just paid me, by order of Lascaridi, cheques for a few hundreds, to meet certain bills endorsed by his firm. My astonishment was not small when he asked for a mortgage.

"What!" I said, "security to *you* for money belonging to Lascaridi and Co., when I know that you owe money to that firm?"

I left him resolved, though a little late, to take the desperate step which could alone save me from impending ruin. I sought George Lascaridi instantly, but he was in Paris. His partners had begun to arrive from the South and the East. I saw his cousin Peter, who seemed both troubled and angry. He was, however, very polite to me. I returned immediately to my office, and wrote a long letter to George Lascaridi, in which I recounted Barker's behaviour. I said I was determined to see his cousin, and come to an explanation with him upon every point, as I was resolved to see what could be done to save both concerns. I received the following reply:—

Paris, Sept. 18th, 1859.

DEAR XENOS,

I found by your letter of yesterday that you have again begun to accuse me. But I suppose you wrote in a moment of passion. If Barker owes money to Lascaridi and Co., that is his own affair; it is not my fault. I did not suggest to him to ask you for security. You do well not to believe it. I see nothing extraordinary in Barker's asking security. He has not a large capital himself; he must look how he turns his business. You say you will have an explanation with Peter. But, my friend, just think that you will expose me. Nothing good can come of it. Why not use all the means we possess to set our business right? I am myself willing to give guarantees. I have done so up to this time, and I have proved it clearly to you. Let us still do the same. The brave man proves his courage in difficulties. I shall be in London next Wednesday. Come to my house at eight in the evening. We shall then consider what is best to be done. To cry out without reflection is not the way to save the company in which I had and have my hopes, after all my other means are exhausted. Good-bye. Persevere if you wish to see the company progress. I am ready to make any sacrifice, because it is for my interest.

GEORGE LASCARIDI.

When Lascaridi arrived in London, he and I had many interviews; many angry words passed between us. However, seeing the chaos in which he was plunged, I abstained from communicating with his partners.

All this time my steamers were at full work, making profitable voyages, having splendid freights. Spite of the heavy interest I was paying, I hoped to be able to clear my vessels. In this my house at Galatz alone could help me. It was on this account solely that I had removed Stockar, and sent out Theologos and Carnegie. But, as we shall presently see, they, instead of assisting, embarrassed me by continually asking for supplies of money, to pay cash for grain for which they had contracted.

I now lost sight of George Lascaridi again. He was like a knight-errant, who, having achieved a Don Quixotic feat in one place, immediately hurried off to an opposite quarter of the globe to tender his services to suffering humanity in that region. It was about this time—November, 1859—that I made a bargain with the Admiralty for the Powerful as a transport for the Chinese War. I made, on that transaction, a net profit of £15,000. This was very well; but I had difficulties to counterbalance my profits. It was about this time that I had to pay instalments on the steamers under construction; and, unfortunately, at the same time several thousand quarters of grain, purchased by my agents, were frost-bound in the Danube, and could not be shipped to England to be realized. Under these circumstances I commenced to negotiate another loan of £20,000 with Messrs. Overend, Gurney, and Co., giving security another steamer. This transaction was also effected through Mr. Henry Barker. As for George Lascaridi, he had disappeared. Poor fellow! he was suffering under domestic misfortunes. I felt deeply for him, but could neither do as he wished, nor as I wished myself. I was bound by honour to my post. The Greek and Oriental was not only solvent, it was coining. To associate that company with Lascaridi and Co., who were standing on the brink of a precipice, would be madness; to make a deed of partnership with George Lascaridi individually, who was involved in so many difficulties, would be to couple insanity with injustice. I stood watching the tide of events, prepared to act as circumstances should suggest. The readers of these pages shall decide whether I was under any obligation to risk the fruit of my hard labours with any one.

One point more. I had come to the resolution of dispensing with the services of Mr. Henry Barker. I resolved to make the personal acquaintance of the mighty capitalists. I was prompted to take this step the more especially because in the approaching January, February, and March of 1860 my payments would, in these three months alone, amount to £143,332 19s. 4d.*

---

* My bills due in January were £30,307 10s.; in February £38,488 5s. 1d.; in March, £74,537 4s. 3d. The bills of the remaining nine months amounted to only £29,383 4s. 10d., as follows—April, £5482; May, £4000; June, 3750; July, £5352 5s.; August, £2250; September, £1750; October, £2750; November, £1750; December, £2300. In these acceptances were included all the above-named mortgages of Messrs. Overend, Gurney, and Co., and others.

CHAPTER XII.

THE MINIATURE COURT OF LOUIS XIV.

It was during the negotiation of this loan of £20,000 that I first made the personal acquaintance of Messrs. Overend, Gurney, and Co.

If characters so different as Adonis, the Chevalier de Faublas, the great Jean Baptiste Colbert, and James Wilson, the eminent economist, could be brought together in one group, they would not form a more strongly contrasting *tableau* than did the four acting partners in this great commercial house.

David Ward Chapman was one of the handsomest Englishmen I have ever seen. He was of the middle height, and perfectly well made. He dressed with the correct taste of a City man, which the "West-end swells" often make ridiculous by exaggeration. He was not a financial genius, but he possessed talent and experience He daily passed five hours in the City. When money was cheap, the commercial sky clear, and business smooth, these hours were, to David Ward Chapman, a sacrifice laid on the altar of duty; but when money was dear, and the aspect of the horizon stormy, then the feelings of David Ward Chapman, during his City hours, were about as pleasant as those of the Syracusan tyrant in presence of the hair-suspended sword. David Ward Chapman was fond of flattery, and, as he could command the market, he had always an abundant supply of the commodity. He was surrounded by a troop of sycophants—all clever fellows—who knew how to trim their sails according to the weather. And then David Ward Chapman was arrogant in the supreme degree. The sentiments of the renowned De Coucy might not inaptly be uttered by him :—

> Roy ne puis-je être ;
> Duc ne veux être,
> Ni comte aussy ;
> Mais grand seigneur De Coucy.

David Ward Chapman loved the drama, and was fond of pictures, but, not being possessed of the judgment of Sir Charles Eastlake, he left the choice of his purchases to his agents or flatterers. In private life he maintained a hospitality lavish as that of the prodigal son. David Ward Chapmân's life was a round of pleasure. With the exception of the few hours daily devoted to business, his whole time was taken up in giving or accepting entertainments. He thought only of making the hours of this ephemeral life as joyous as possible. Prodigal-son like, he never reflected on the reverses of Polycrates.

Arthur George Chapman was a fair-complexioned, boyish-looking young man. Not a particle of hair shadowed his downy cheek. Had he chosen to dress like a girl, he, like the Chevalier de Faublas, might have passed as *femme de chambre* of the Marquise de B———. The part he played in the great office was limited to issuing orders in a bombastic tone to the clerks, in the presence of strangers; to treating customers cavalierly, particularly those who came to borrow money, or ask the renewal of their bills. But, over and above all, his object in coming to the office was to accept the bills of the firm and sign cheques. Indeed he signed so many cheques, that within eighteen months from the time I first made his acquaintance, I was told, his partners, to relieve him from this heavy work, presented him with an annuity. So ended his official visits to the great "Corner House."

Henry Edmund Gurney was one of the principal partners and capitalists in the establishment. He was a man of a high moral and intellectual cast. In person he was above the middle height, and inclined to corpulency. His countenance indicated both goodness of heart and strength of intellect. His hair was fair, his eyes well-shaped and kindly in expression. He was deaf in one ear, which accounted for his always speaking in a loud tone, even when treating of affairs of the most private nature. He was a proud-spirited man, somewhat pompous and dictatorial in manner; but, on the other hand, he was imbued with a deep sense of religion, and his acts of private benevolence were numerous. In his public capacity he has, many a time and oft, saved whole branches of commerce from a crash, by assisting those at their head. He was a member of the Society of Friends, and a man incapable of uttering an untruth, or playing a double game. But, unfortunately, surrounded by a class of men like the favourites of the Chapmans, he was not always able to ascertain the truth, because, not suspecting the motives of his informants, he never thought of doubting their statements. In this way he ultimately fell into their power, and became the tool of a clique of designing men. Like all financiers, Henry Edmund Gurney was fond of multiplying his money, and was ambitious of commanding the commercial world. It was an ambition worthy a minister of finance. To reach the climax of his ambition, this English Colbert did not need the patronage of a Cardinal Mazarin.

When money was cheap his eye instantly fell on the dockyards of Liverpool, London, Hartlepool, Newcastle, and Sunderland, as did, in former times, that of the celebrated minister of France on those of Brest, Toulon, and Rochefort.

With the vast capital at his command—partly his own, and partly deposited in his keeping—he might, by mortgages, have chained the mercantile fleets of England, and compelled them to anchor where he pleased. It was to a power like this that Henry Edmund Gurney aspired. Delille has said: "*Par les nœuds du commerce unissez l'univers.*" Henry Edmund Gurney's plan was not a bad one. Instead of incurring risk by discounting second-class bills, as all bill-discounters do when money is cheap, he was able to put his hand upon a vast amount of commercial and maritime property, and so, in a large measure, to control the two chief elements of England's commercial greatness. To the success of Henry Edmund Gurney's design three conditions were indispensable—a special knowledge of maritime property; a moderate rate of interest, say, never in any case to exceed 10 per cent., so that the shipowner may be able gradually to release his vessels; and an inviolable rule never to advance more than a certain amount on a certain valuation. Besides, the trade in which the mortgaged vessels were engaged should be closely inquired into, and the capabilities and honesty of their owners ascertained. But the mode in which our great capitalists attempted to work out their project was quite the reverse of this. The *protégés* whom David Ward Chapman employed to superintend their interests and to value the property, were babes in knowledge, or else refined rogues, and the consequence was that transactions with the best paying and, best organized lines of steamers were fraught with loss. Mr. Henry Edmund Gurney, being the chief acting partner in the firm, having, unfortunately, more on his hands than one man could accomplish, left the investigation of the securities to Mr. David Ward Chapman, whose factotum was Mr. Edward Watkin Edwards, an official assignee of the Court of Bankruptcy, and whose acquaintance we shall make shortly.

Robert Birkbeck, a junior partner in the great "Corner House," was at that time, like Arthur Chapman, a boyish-looking young man. He possessed great talent, prudence, and high morality. He loved his work, and took an active part in all the concerns of the firm. Endowed with great self-possession, he stood unmoved amidst commercial storms, and more than once, by his foresight and determination, saved the good ship of Lombard Street from being wrecked amidst the labyrinth of shoals and banks amidst which the two other partners and their precious favourites had entangled her. Robert Birkbeck stood unflinchingly by the side of Henry Edmund Gurney to the last, and, I doubt not, rendered good service. If all the winds of the commercial horizon of London had not been evoked and made to blow simultaneously on that glorious ship, which, after I left, became a total wreck, she might have lasted to become an honour to Great Britain.

Samuel Gurney and John Henry Gurney scarcely ever made their appearance at that time in the office.

The way in which Messrs. Overend, Gurney, and Co. advanced money to the Galway Company, to me, to Zachariah Pearson, and other shipowners, on their property, was this: They instructed Mr. Henry John Barker, or the introducer of the business, to draw bills on the owner of the property, and these they discounted at the enormous rates I have mentioned. These bills they kept by them, and re-discounted, with their indorsement, according to the signals indicating whence the money-market winds blew, or according to the wants of their claimants.

In my visits to the office of these great moneydealers, I had an opportunity of noting things that escaped the observation of the proprietors themselves. That office often reminded me of the French court in the time of Louis XIV. "*C'est à la cour que les passions s'excitent, et conspirent contre l'innocence*," says Fléchier; and Arnauld adds, "*Ici la fourberie passe pour la vertu*." The truth of these assertions was fully verified in the miniature court of Lombard Street. No ante-room of any minister in Christendom presented a greater number of expectants than was frequently to be found there from ten in the morning till four in the afternoon. Persons of all grades, from the high to the low, were to be seen waiting for an audience, or even a reply through a clerk. You saw there the "*peste de cour*," as Molière calls those officious insignificant courtiers whose sole occupation is to circulate groundless rumours and calumnies. It was amusing to see specimens of that class at the Lombard Street court, and observe the sycophantish smile and slavish humility with which they approached a disengaged partner, and commenced an under-toned gossip. You saw there the "*renard de cour*," as Balzac calls the crafty courtier. He approaches the great man with a cautious and insinuating air, and suggests—merely throws out a hint about a promising transaction, but he does not venture to propose anything until he has sounded the great man's inclinations. Then there was the "*mouche de cour*," as La Bruyère calls the professional *espion courtisan*, for his reliable information, that might either save the great house from loss, or bring it profit. These men were well paid by commission, or in some other way. Then there were the flocks of commonplace courtiers, made up of brokers and *proxenites* of all kinds, full of all sorts of schemes and propositions. These went hither and thither, flinging the *eau bénite de la cour*, or the vain hopes and promises, to their anxious clients. It was of these last that Corneille said:

> Les courtisans sont des jetons ;
> Leur valeur dépend de leur place :
> Dans la faveur des millions,
> Et des zéros dans la disgrace.

What plots have I known to be concocted, what intrigues have I seen played out to supplant a favourite in that little Lombard Street court! Thomas Corneille has sublimely said:

> Le crime fait la honte, et non pas l'échafaud.

This is no doubt true in the abstract, and will be proved correct when the universe, as a whole, is brought to account; but, in the brief span of human existence, we constantly see the converse of the assertion pass for truth. Many an honest and honourable man have I seen driven, by the corrupt favourites of that house, to the scaffold of bankruptcy, and, though innocent, covered with ignominy and that shame the world still attaches to their name. Whilst I write I have "in my mind's eye" Zachariah Pearson, of Hull, as honourable a man as any in England, who was thus sacrificed.

The " Corner House," in Lombard Street, consisted of several floors, and had an imposing appearance from the outside. It was only on the ground floor that business was transacted. On this floor was the large room for the clerks, which had the appearance of a bank, though the clerks were not so numerous. Next to this room was that in which the partners sat, and into which entrance was gained from the clerks' room through a glass door. This room, in which used to sit the greatest money-dealers in the world, was plainly, almost meanly furnished. There were in it five desks, ranged one after the other, on the plan adopted in the Lancashire schools; these desks were fenced round with a breast-high partition. At these desks some of the partners were always to be found writing, but Mr. Henry Edmund Gurney was generally standing by the chimney-piece, attending to business. I may say that, for a commercial firm of such magnitude, the furniture and general appearance of this room were rather shabby. There was not even a spark of that ostentation which characterizes the French *comptoirs* and the newly-built banks of London. Next to the partners' room were two other small rooms—one looking into Birchin Lane, pretty respectable as regarded light and furniture, the other almost as grim and dark as a ship's lazaret. Happy the thoughts of the man who had to kick his heels in this room for the best part of an hour, whilst waiting for an audience, surrounded by half-empty medicine bottles, brushes, combs, towel-stand, and washing apparatus. Beyond these was another room—light, airy, and comfortable—communicating with the partners' room through the clerks' room.

CHAPTER XIII.

A PERILOUS POSITION.

As soon as I became thoroughly acquainted with the politics of the leviathan money-dealers, and the system that obtained in their office, I came to the conclusion that the less frequent my visits were the better.

I have mentioned how I accepted bills for above £100,000 against the grain and steamers I had bought. These bills now began to fall thick. As the Danube business and the import grain were not credit transactions, but required hard cash, my position became alarming. This, Mr. Henry John Barker was the first to perceive, and, under pretext of being extremely busy, he avoided meeting me. George Lascaridi was more erratic than ever. Rapid in his transitions as one of the Dioscuri, he was to be seen one week in Paris and the next in London. I was in a dilemma. Within ten days I had to meet £38,000, and I did not know even where to look for the money, though I had still plenty of ship property untrammelled. I went home that evening in a most sorrowful mood. An idea suddenly crossed my mind. I sat down and wrote a very long letter to Henry Edmund Gurney. I explained the exact state of this line. I showed that my business was yielding a profit of at least 30 per cent., and that it was based on a good foundation. I only needed at the existing crisis a small capital of £80,000. If that were advanced, I would willingly give security, and would, moreover, give one-eighth of the yearly profits of the business, as long as it existed, after repaying the money lent. As this letter is the real medium of the introduction of the Greek and Oriental Steam Navigation Company to Messrs. Overend, Gurney, and Co., I publish its rough copy, which I found amongst my papers. The beginning of this important letter, bearing the date (which can be but a very few lines), I have lost.

We had beaten opponents and driven them out of the way, causing, perhaps, a great deal of jealousy, and we had in reality a field of gold before us, when Mr. Lascaridi unfortunately, and without my knowledge in the least, mixed himself up in the Galway Company on an enormous scale. How he did so, and on what prospects, is till now a mystery to me. When I learnt it, I foresaw at once that the *contre-coup* of the Galway Company would fall immediately on me. I remonstrated with Mr. Lascaridi most sincerely, and I proved to him long ago that this affair would shake the credit of his house. It was too late, and Mr. Lascaridi had full confidence in the promises and prospects of Mr. Lever. However, he did not pass one month, and Mr. Lascaridi could not give me either his moral or his pecuniary assistance. The contrary; the more his credit suffered, I suffered too, and at last he left all the tremendous weight of the company on my shoulders and on my resources. From that time I began to mortgage the steamers to you to carry out the company, to save my honour, my position, and principally not to disappoint the shippers that had deserted others to come to me.

Liabilities.—The company, true, makes money fast, and paid under two years, out of the profits—who would believe it?—nearly the half of the cost value of our steamers, but has still nearly £130,000 to pay from the 1st of this year to the July of 1861; your loans, amounting to a total of £52,000 (deducting the loss of the Tzamados); and £28,000 to Messrs. Pickersgill, Rodocanachi, and the Bank of London. Against that I hold eighteen new steamers, cost value more than £250,000; I hold 15,000 tons of coal in the different dépôts of the Mediterranean, at the low price of 30 s. per ton, £21,000; and 6000 quarters of grain, value nearly £10,000; and uncollected freights from the Government, £70,000, of the Powerful, Asia, and Scotia,* for the expedition to China, at the end of eighteen months. And it is expected the other steamers, now with the new Bank of Turkey, when the Exchange shall be established at Constantinople, our exportation will increase, so that they will do better.

Relations of the Company with others.—The company has no relation with any one, except with the house of Messrs. Lascaridi and Co. No failure can injure the company—even the house of Messrs. Lascaridi and Co.—because we do not hold any acceptances of them; but if we fail, we shall affect them, because their name is endorsed on our acceptances.

Its Solidity.—To me—I that have the sole direction of the company, and must know everything regarding the company—I consider that the company, according to the best of my knowledge, is as solid as any business in London, but pressed very hard for money, and wanting the necessary funds to work it respectably, because we have to pay in cash immediately for everything—all the wages, provisions, and stores—and meet the bills of the builders, when, on the other part, we cannot collect the freights till more than three weeks after discharging. I repeat, I can prove to you by our books the company is solid, and whatever crush or undervaluation to her property may be made, it will be found at least 30s. to the pound.

After all this long tale, I come to the principal object that made me write you this letter. My object is this, only this—to protect your interests and the money you have trusted to me, and to prevent your name from being the object of public notoriety, if, unfortunately, I cannot pull through.

I shall not do justice to myself if I do not tell you clearly the position in which I am at this moment with Mr. Lascaridi and Mr. Barker. I said that we have £130,000 liabilities—viz., bills due to the ship-builders, and other transactions. If these bills were to be met in the course of the eighteen months, the affair would be done very comfortably, working the steamers; but the principal amount, £80,000, is due in this month, February, and March, and we have no other means to meet it but by accommodation, and such accommodation, with such bad credit as we have,

---

* The Asia and Scotia were not taken up by the Government, being at that time in the Mediterranean, and not arriving in London in time to be inspected as transports.

will be more ruinous than if I sell our property at auction. Such accommodation would ruin our credit more without curing it, because the difficulties of these three months, if they will be over, will pass to the next three months, and, having to make sacrifices, pay heavy interests for meeting the payments of the day. I really must add that I shall be, as I am already, in perpetual agony. I shall never be able to spare anything to pay you to release my mortgaged steamers, and, worst of all, I shall neglect the business, and the transactions of the office, by being always out of doors to find money. The end will be a very bad case for all parties interested in this business.

After long reflection, without saying anything to Barker or Lascaridi, I decided to follow my conscience and the path of honour, and I do not care if I remain without a penny in the world. There are two ways—either to apply to you for a further loan of £80,000, giving you security, so as to take up our acceptances, and then every quarter I shall be able, from the profits, to pay you off at least £10,000 of this loan; or to suspend payments, and to ask for time to pay everybody 20s. to the pound. This, I know, is ruinous, and this beautiful business will go to smash, and it will raise against us, and against the Greeks a great deal of scandal; but it is a hundred times better than to drive the business any further in the way that we are going, which time, instead of curing, will prove more fatal.

Asking you for £80,000.—I propose certain conditions that you will find not only profitable, but at the same time inspiring more confidence.

First.—I shall take out of Messrs. Pickersgill and Rodocanachi and the Bank of London my three best steamers—viz., the Scotia, value £25,000; the Modern Greek, value paid £20,000; and the Powerful, value £42,000, including the £28,000, her freight from the Government, giving you an order monthly to collect it, and mortgage them to you, and then you will be the only mortgagee of our company. Those steamers I was obliged to mortgage for a few days for a very few thousand pounds; they are included in the above liability of £130,000.

I give you, in addition, the Smyrna, value £10,000; the Patras, value £8000; and the Colleti, value £11,000. They are free, and the Petro Bey just launched.

Secondly.—When the Scotia and Asia return to London (now

on their passage from the Black Sea), and the contract is signed by the Government for transports, as arranged with the surveyors, I give you an order to collect monthly also £28,000 on each vessel's freight; and, by the three freights only of the Government, you will have back £73,000 of your money.

Thirdly.—I give you a written promise not to build or buy any steamer before I take out of your hands the mortgages of all our steamers.

Fourthly.—Not to accept for any account any more bills, except for the value of corn drawn from the Danube. I have no objection, if you send me one of your confidential clerks, to place him in my booking department, so as to see how things work.

Fifthly.—The interest of the money advanced to us to be at the rate of 10 per cent. per annum; but, when I redeem the property —which I expect will be in not longer than eighteen months—I propose to give you by contract, as long as the company exists, one-eighth of its profits annually; or, if you choose, to register at once in your name the eighth part of every steamer, or to give you a bill of sale for the eighth, and then you will be entitled to the eighth part of the profits that the steamer makes.

If you find my proposition a little liberal, and entertain it, then I shall explain to you why I do not apply to the numerous friends that I have among the Greek houses, or why I do not make the company a public company. There are very few houses that can advance me such a large amount. I shall be obliged to give them all the above explanations—to show them my profits. I shall open their eyes; and the result will be that, being exporters and importers themselves on a large scale, having all the means, they will try the business, and will go themselves into the shipowning—as already Messrs. Spartali and Co. have begun at Liverpool—the Greeks unfortunately following each other in a transaction; so, in a couple of years, every two or three houses will have three or four steamers, and then it will be the same competition and the same losses as to the lines of Holland, Belgium, Hamburg, and America. In fact every one would have to work for a 5 per cent. instead of a 40 or 50 per cent.; so, as long as I can keep alive the Greek and Oriental, I must keep in darkness those people from whom I only fear a true and ruinous competition. The same reason compels me not to make the company a public company, although Mr. Lascaridi pressed me a great deal. If I make it public, the greater part of the shares must be sold only to the shippers—viz., to the Greeks—to secure their business and support. The directors must be the most part Greeks; immediately they will start another company—the business will be divided; it will be a competition, and, if once competition is multiplied, each company soon will come to an end. This business will not bear competition, because competition is equal to ruin. Another thing: once this company has a board and many directors, it will go to pieces, because, when one director will think that barley is the best cargo to ship home, the other will say wheat or Indian corn. The steamer will remain at Constantinople, waiting telegraphic orders, till the board meets and decides—perhaps twelve or fifteen days. Another thing: a steam company is only for mails and passengers—it has its route traced forwards and backwards. In other words, if till now it existed a commercial limited house, to import corn, with directors and board, then this may be the same. I understand one, or two, or three partners being all day in the same office, and, knowing everything, to act accordingly; but many directors, having other business, and attending once every week, experience teaches me that such company will not last long. It will be a clear and net imposition on the public, or, better saying, a clear robbery on the part of the promoters or directors.*

You will tell me it is better to do that than to leave the company to be smashed if we do not help you. Permit me to think differently. If the Greek and Oriental is smashed now, no one has to lose money, even if my property is sold for half its value. I am not to be blamed, true—cruel position!—but the public will see what I constructed with only my little means; what strict economy I kept in the office, in the ships, and in my private life. I shall come out, perhaps, poor, but I am certain with great credit and honour, and I have no doubt that the creditors will give me the only thing I want now—time. To form the company limited at this moment is also impossible. If I overcome these three months, I shall get out altogether shortly, and with great property and a good business. I shall divide always my last penny with Mr. Lascaridi. Mr. Lascaridi had and has all the good will to give me any assistance in his power. He is the most honourable man that I met; but, unfortunately, he met with people who abused his good

---

* What I wrote here in January, 1860, was proved in 1866 in the Anglo-Greek Steam Navigation and Trading Company, Limited.

will. He has suffered—his money is locked up, and he cannot now assist me. That is not a reason to abandon him. He said to me many times that, if he knew my business was so good, he never would do any other thing than devote his money and time to it. However, with all this, in this moment I must follow my own head as to what is to be done—what is the best for all—and really I do not see any other way more clear and honest than to leave *à part* timidity and shame, and to tell you all, and to ask you, giving you security and benefit, to help us to get out of this labyrinth. Doing so, you help not only me, and Mr. Lascaridi, and Mr. Barker, but you do a great deal of good to nearly 250 English families that live out of our steamers.

Our company made at least a great deal of good in the sugar and colonial markets, and in the corn market, so its existence will be a great honour to who can save it.

This is, dear Sir, my sincere confession. I have told you all, and I beg you to show my letter to Mr. Chapman, and at least, if you refuse my proposition, to advise me what you think the best course to follow.

Yours truly,
(Signed)   STEFANOS XENOS.

CHAPTER XIV.

MESSRS. OVEREND, GURNEY, AND CO.

MONEY was at that time very cheap. Messrs. Overend, Gurney, and Co., after a short consultation, decided upon advancing the sum I asked. I was to pay interest and commission, but they declined participating in the profits of my business, as that might establish a kind of partnership between them and me. Mr. Henry John Barker and Mr. George Lascaridi, those two gentlemen who had been missing from the circle of my acquaintance for some time, now suddenly reappeared within it, with countenances so smiling that it was a real pleasure to look on them.

Mr. David Ward Chapman informed me that Messrs. Overend, Gurney, and Co. had appointed Mr. Edward Watkin Edwards, official assignee in the Court of Bankruptcy, as the gentleman who was to look after their interests in my business, in accordance with the spirit of the letter which I had written to Mr. Henry Edmund Gurney. Mr. Chapman recommended me to go at once to Mr. Edwards's office, in Weaver's Hall, Basinghall Street, in company with Mr. Henry John Barker, who was present at the interview, and see him (Mr. Edwards) on the subject. There being no time to lose, we started at once. When we arrived at Weaver's Hall we found that Mr. Edwards was engaged, and would remain so for a short time. In a few minutes there emerged from Mr. Edwards's private room an elderly man, rather shabbily dressed. He was growling and swearing, most bitterly, and speaking of Mr. Edwards in anything but a complimentary manner. This incident made a most painful impression on me. On our being announced, Mr. Edwards, who had been already apprised of the object of our visit by Mr. Chapman, received us with a smiling countenance; but I was not slow to discern the traces of strong agitation, caused by his recent visitor, with regard to whom he made the remark, that he had been very well off prior to his failure; that his accounts were in great disorder; that he was mad, and did not know what he was saying. I do not know why, but my first visit to Mr. Edwards gave me an unfavourable impression regarding him; yet he was a man of pleasing appearance, placid countenance, cool temper, of not only gentlemanly, but fascinating manner, and soft and sweet of speech.

I explained to Mr. Edwards the cause of my visit, the state of my affairs, the nature of my business, and the number and value of my steamers. He noted what I said on a paper which he had before him, and, when I had finished, told me that Mr. Chāpman had consulted him about the matter, and that he, seeing that the business was really a good one, had advised the firm to make the advance asked; all, therefore, that was to be done to finish the transaction was for me to meet him (Mr. Edwards), at Messrs. Overend, Gurney, and Co.'s, in half an hour. Punctually to the time appointed I reached the "Corner House," in company with Mr. Henry John Barker. Mr. Henry Edmund Gurney was out of town that day. When we entered we found Mr. Edwards engaged in a private conversation with Mr. David Ward Chapman,

in one of the small rooms. After some little time
Mr. Edwards called us in, and then he informed me
that Messrs. Overend, Gurney, and Co. had decided
on advancing me £80,000 for six months, to enable
me to overcome my difficulties; but for that advance
I should have to pay them a bonus of FORTY THOUSAND
POUNDS, and interest at the rate of 10 per cent.

"What! £40,000 and 10 per cent. interest?" I
screamed rather than asked.

"This is cheaper, after all, than the eighth of
your profits, which you proposed yourself to give us
every year," said Mr. Chapman, with an air of great
indifference.

"It may be so," I answered; "but to pay you
£40,000 and 10 per cent. interest in one sum at once,
with £80,000 in six months, is impossible; I could
not afford it; I should never be able to do it; it is out
of the question, Mr. Chapman."

Mr. Edwards directed a significant look towards
Mr. Chapman, and then said to him, "Just one word
with you," taking him at the same time out of the
room.

"Forty thousand pounds bonus and 10 per cent.
interest besides, all to be paid in six months! I
would rather stop payment at once. It is too much,"
said I.

"Yes; it is too much," said Mr. Barker, but not
in the same surprised manner that I had done. He
looked to me as if he was familiar with such transactions.

After the lapse of a few minutes the official
assignee returned with a beaming countenance. "I
have arranged your affair satisfactorily," said he;
"I have persuaded them to accept a bonus of only

£30,000, with 5 per cent. per annum on the money they will advance you, and I have no doubt but that, if at the end of six months they find your business going on all right, they will renew the loan for another six months, at the same rate of interest, and without any bonus. Now, do not throw impediments in the way, and you will be all right shortly. Barker Brothers (which was the name of Mr. Henry John Barker's firm) will draw a bill on you for the total amount, which you will accept; that bill will never leave this office, so that you will be quite comfortable."

I had no time to lose, neither had I cause to complain—the choice was between life and death.

" When can I have the money," I asked; " as I have to meet £30,000 worth of bills this week ? "

" You can have it to-day, if you like," he replied.

I said, " Very well; I suppose I must accept. The terms are rather stiff."

Mr. Chapman was then called in, and seemed much pleased on hearing that I had accepted their terms. Mr. Edwards then told him that I must have a cheque for £20,000 at once. " Very well," said Mr. Chapman; " but let us give our cheque to Barker Brothers, and Barker Brothers will give you theirs. We do not wish the name of the Greek and Oriental Steam Navigation Company to appear in our books. This does not make any difference to you so long as you receive the money. We will give you the remainder to-morrow." The cheque was handed to Mr. Barker, and we left the " Corner House " together, to get his firm's cheque. My brain was nearly stunned by what I had seen and heard.

" What a fool;" cried Mr. Barker, striking his forehead with his open hand when we had got into the street. " Had I only known that they were going to charge you £30,000 for bonus, I would have made them give me £5000 out of what they are going to receive. An opportunity of getting such a round sum out of them seldom occurs."

"You ought certainly to ask them to give you £5000," said I, astonished at the extravagance of his ideas. I had to work like a horse to make my poor steamers pay—nobody can guess how hard I had to work—and yet Mr. Barker wanted £5,000 for walking with me from Lombard Street to Basinghall Street and back, and thence to Abchurch Lane. Payment as princely as this would make it worth the while of the son of a duke to come to the City. Perhaps it was such examples that caused so great an influx of noblemen, admirals, and generals into the City during the time of the limited liability companies. Be that as it may, Mr. Barker gave me his cheque, and I returned to my office.

CHAPTER XV.

MY FIRST FINANCER.

MR. EDWARDS was regarded at the "Corner House" as a great mathematician and a high financial authority. He was a barrister by education, and an official assignee by profession. If the commercial laws of this country were not so complicated, surely the last person that two merchants ought to invite to step in to arrange their affairs would be a lawyer. However, Messrs. Overend, Gurney, and Co., upon principle, kept in their employment this gentleman, introduced to them by Mr. David Ward Chapman, hoping that his legal and shipping knowledge might avert complications. This semi-professional gentleman led them into troubles, and got them out of their difficulties by sinking them each time deeper and deeper.

After the lapse of a few weeks, I saw that Mr. Edwards had no intention of coming to my office. He had advised the house to advance me the sum of £80,000, without even going himself or sending some competent person to inspect the steamers, a mortgage on which was to be the security for the amount lent. I was amazed at the facility with which this gigantic transaction was completed in the short space of eight-and-forty hours. Mr. Edwards, to whose name, as the nominee of Messrs. Overend, Gurney, and Co., the steamers were to be mortgaged, did not even ask me, till several months after, for the policies of insurance on them; and it was a week after I had all the money that I signed the bills of mortgage. In fact, I received nearly all this large sum of money without having given any security in return.

This mode of doing business caused me to reflect most seriously. I said to myself, " When a banking-house, no matter how great its wealth, transacts business in this manner, it must surely have very important reasons for doing so." For my guidance in future, I worried my brains to discover these reasons, as the security I had to give was larger than the amount advanced. "Is it," I said, "that Messrs. Overend, Gurney, and Co. have unlimited confidence in me?" But that could not be, as I was almost an entire stranger to them, and a foreigner besides. "Is it because the partners are ignorant of the conduct and incompetency of the gentlemen appointed to super-intend their interests?" It could not be that, for he was an old acquaintance of David Ward Chapman's. "Is it because the colossal bill of £170,000 * which I am to accept is to be drawn by Messrs. Lascaridi and Co., according to the request of Messrs. Overend, Gurney, and Co.?" But Messrs. Overend, Gurney, and Co. knew that Lascaridi and Co. were not solid, owing to the enormous fabric of accommodation bills built up between them and their representative, Mr. Edwards, in order to "finance," as he called it, the famous Galway Company. My evil thoughts even went so far as to imagine that this great banking-house was rotten, after all, and that those who were

---

* This bill was made £170,000, because the former mortgages were included in it.

acquainted with the secret were profiting by their knowledge of the existing confusion. This supposition might have become a settled conviction had it not been dispelled by a certain eventful conversation.

When I was directed by Messrs. Overend, Gurney, and Co. to go to Mr. Edwards, in order to complete the negotiations for this loan, the official assignee asked me, in plain English, how much I would allow him annually for his trouble. I replied that he must look to his employers for payment, and added that they could afford to do so out of the £30,000 bonus and 5 per cent. interest I was to pay them for the loan. He said that he had spoken to them about it, and that they told him he must look to me for remuneration. He added that he would be satisfied with £500 a year. I did not know then whether Messrs. Overend, Gurney, and Co. authorized him to ask for this annuity from me, or whether he received commissions without their knowledge, or with the knowledge of Mr. David Ward Chapman alone, with whom he was most intimate, and associating daily at Brighton, where they were both residing. This, however, I do know, that I was forced to accept his terms in order to have the loan completed. This conversation dispelled the idea of insolvency, for I naturally concluded that Mr. Edwards, who was the "law and the prophets" of the firm in all their transactions, would have known how they stood, and, instead of asking for an annuity, would have insisted on a lump sum once for all out of the money advanced through him to me.

In all this transaction there was something most inexplicable to me. It seemed as if the great firm were possessed of some secret, and that I was to be initiated in some Samo-Thracian mysteries; but, as I was still only a neophyte, I could not pass to the highest degree until I had been fully catechized and taught, and promoted from order to order.

When I first negotiated this loan, I did it direct with Mr. Henry Edmund Gurney. It was the result of the private letter which I wrote him, and the security was to be a mortgage on several of my steamers and freights. After I had had a portion of the money, Mr. Edwards and Mr. David Ward Chapman announced to me that the bill for £170,000, which I should have to give Messrs. Overend, Gurney, and Co., would be drawn on the Greek and Oriental Steam Navigation Company by Messrs. Lascaridi and Co. I had no objection, as it did not matter an iota to me who drew it. But here again I was provided with food for reflection. At that time Mr. George Lascaridi was, as I have said, the managing partner, in London, of the Greek firm of Lascaridi and Co.; the other and senior partners were at Constantinople and Marseilles. Was it with their consent and knowledge that he pledged the signature of his firm to Messrs. Overend, Gurney, and Co., for the enormous sum of £170,000— a sum larger than their whole capital? That could not be, because he had repeatedly told me that his partners would have nothing to do with my steamers. Was it for a commission? Was it to oblige Edwards and Chapman, and to throw dust in the eyes of the Gurneys and Birkbeck? Was it because he was the capitalist of Mr. Henry John Barker's bill-discounting office in Abchurch Lane, or was it because he was connected with Mr. Edward Watkin Edwards, the official assignee, in financing the Galway Company? When I asked him afterwards why he signed the name of his firm for such an enormous amount, he replied that he could not help it, as he had already endorsed several bills of the Greek and Oriental Steam Navigation Company, which we had given to the ship-builders. Any one who has read the commencement of this book will know how I formed that company, and for what consideration the indorsement of Lascaridi and Co. was given. Therefore, if Mr. George Lascaridi really wished to rid himself of his firm's liabilities on account of the Greek and Oriental, here was the opportunity—to leave Barker Brothers to draw and endorse the bills; for the money I was borrowing from Messrs. Overend, Gurney, and Co. was to be used for the payment of the bills of exchange, bearing the indorsement of Lascaridi and Co., which had been given to the ship-builders.

There is one class of men who consider themselves very far-sighted, but who will, nevertheless, enter into speculations without pausing to think of their consequences; and there is another who will not make a step in advance without stopping to examine carefully the ground over which they are travelling. Had I belonged to the first class, I should now be what Mr. David Ward Chapman once told me I should become, owing to my profitable business, and to my connexion with them—one of the richest men in the City of London.

The next day and the day after that on which this colossal transaction had been almost completed, I awaited in vain the advent of Mr. Edwards, who had been appointed by Messrs. Overend, Gurney, and Co. to superintend my finances, in my office; but he never came. I had mistaken him. He was by far too high a personage to trouble himself about so small a matter as £170,000; he had other more important duties to perform. When one day I told him that I had been expecting a visit from him, he answered me by saying, "I have no time. I am overloaded with Messrs. Overend, Gurney, and Co.'s business, Xenos; I shall therefore send my confidential clerk, Mr. March, every evening. He will go through the books with your book-keeper, and then strike a balance-sheet which will show us how you stand. It is just the same thing as if I went myself, for Mr. March is entirely in my confidence, and knows everything."

I said, "Very well; but surely you will give us a call to see the offices and how we do business there?"

"Of course," he answered, "I shall come the moment I have time."

Mr. March called the next day, when I introduced him to Mr. Ross, the book-keeper of the Greek and Oriental Steam Navigation Company. Thenceforth, every evening, after the other clerks had left, they set to work to produce the balance-sheet, which was fair copied by Mr. March, and is now in my possession. It consists of several folios, and its recapitulation is given on the next page.

BALANCE-SHEET, 1859.

Dr.

Dec. 31.
			£	s.	d.
To Creditors unsecured	..p. 4		£40,624	0	8
,, on Bills	..p. 5		77,765	4	11
,, ,, holding Security	..p. 6		112,544	5	5
,, Profits on Trading	..p. 12		64,792	16	9
			£295,726	7	9

Cr.

Dec. 31.
			£	s.	d.
By Cash at Bankers	..p. 1		£484	2	11
,, Debtors	..p. 9		19,870	16	5
,, Property	..p. 9		73,543	16	9
,, ,, held by Creditors	..p. 7		176,889	9	2
,, Interest	..p. 13		4,136	14	7
,, Commission	..p. 13		8,820	1	3
,, Sundry Charges	..p. 14		7,245	1	2
,, Deficiency, Jan. 1, 1859	..	..	4,652	2	4
,, Difference in Posting	..	..	83	18	2
			£295,726	7	9

CHAPTER XVI.

MY SECOND FINANCER.

WITHIN a few months after the completion of the mortgages, viz., the 1st of August, 1860, I was sitting in my office, when Mr. Edward Watkin Edwards, for the first time, entered. He informed me that Mr. Henry Edmund Gurney, fully satisfied as to the value of my business, and with the balance-sheet of Mr. March, wished, as the future capitalist of the Greek and Oriental Steam Navigation Company, that the concern should be turned into a *bonâ fide* private company, and to do that he proposed that I should enter into a deed of partnership with Mr. George Lascaridi, who had endorsed the bills. He added that Mr. Lascaridi was to be put at the head of the financial department of the Greek and Oriental, and was to go every day to Lombard Street, where he could get as much money as he wanted. Mr. Edwards urged me not to make any objection; firstly, because the proposition came from Mr. Henry Edmund Gurney; and secondly, because he was himself so overloaded with business that he could not undertake to superintend mine. And then Messrs. Overend, Gurney, and Co. had been acquainted personally with Mr. George Lascaridi, and could not make a better choice of a partner for me than one of the firm of Lascaridi and Co., and a compatriot of my own.

Having said all this, Mr. Edwards drew from his pocket a preliminary draft of the deed of partnership, which he asked me to sign. He also asked me to sign a letter, engaging myself to pay him, Mr. Edwards, annually, the sum of £500, in consideration of his past and future services. I was taken by surprise. I felt that I was muzzled. Mr. Henry Edmund Gurney's wish was virtually a command. I was fully aware that Mr. Edwards was Mr. David Ward Chapman's bosom companion and prime favourite. I saw that Lascaridi had worked the official assignee. I was caught in the hunter's toils. I said to Mr. Edwards: "I thought Messrs. Overend, Gurney, and Co., by making me pay the bonus of £30,000 for the loan, had given up all pretension of participating in the profits of my business, in accordance with the spirit of the private letters which I had written to Mr. Henry Edmund Gurney. I have given them a bill of exchange for £170,000; when I pay that, they are bound to release my steamers, and I am under no further obligations to them. All they can demand from me now is that you, Mr. Edwards, their nominee, and in whose name my steamers are mortgaged at the Custom-House, or your representative or clerk, may have the power to come into my office to inspect the books and see that the business is going on all right, and for doing this I have already agreed to pay you £500." Mr. Edwards, who had been looking me fixedly in the face during the time I was speaking, at last said:

"Have you seen Lascaridi?"

"Yes, I have," I answered.

"What did he say to you?"

"He said that as my business seemed established now on so solid a basis, as I had succeeded in getting Messrs. Overend, Gurney, and Co. to back me, he would like to have his share in it. He then insinuated that he had a right to his share, as the indorser of the bills of exchange which had been given to the ship-builders, and as the indorser of the big bill for £170,000. I told him that his firm had endorsed the bills on commission, and that if any one was entitled to share the profits of my business it was his firm, to which I had no objection, but as long as his firms in London and Constantinople brought claims against the Greek and Oriental Steam Navigation Company, and held the deposits of coal, I could not enter into any agreement with him. We at last came to high words, and so I left him."

The official assignee listened to me patiently, and then, in that smooth cool manner which is so characteristic of him, said: "Look here, Xenos; you may congratulate yourself on being the most fortunate man in England, for having fallen into the hands of such a firm as the Gurneys. They like you, and now they have taken you by the hand they will make a great man of you. I cannot tell you more at present. They have great plans regarding the Greek and Oriental Steam Navigation Company, but they cannot put them into operation until they have placed the company on a different basis. Do not, then, refuse Mr. Henry Edmund Gurney's request, who has been so good to you, and who saved you from bankruptcy."

"I have no objection to give Mr. George Lascaridi a share in my business, Mr. Edwards," I replied. "I like him; I owe him immense gratitude; he is a distant relative of mine, and I regret now that I had angry words with him yesterday. I do not, however, like partnership; it is like matrimony—if you get a good wife, you are blessed and happy; if you get a bad one, you are ruined for life."

I also observed that Mr. George Lascaridi was

connected, besides the firm of Lascaridi and Co., with other concerns, of whose solvency I knew nothing. Mr. Edwards replied that all Lascaridi's concerns were known to Messrs. Overend, Gurney, and Co., that he was a great favourite with them, and that a partnership with him would give them great confidence in me. Mr. Edwards spoke plausibly. I was conscious of the difficulty of my position.

The mortgagee of maritime property has more power than the mortgagee of landed property. At any time he likes he can give a bill of sale to whomsoever he pleases, transfer your steamers to the name of another, deprive you of your ownership, and take from you all control over your captain and crew. If you have any inkling of his intentions, you can, it is true, prevent him by an injunction in Chancery; if not, your only remedy is at common law, which is a most tedious and expensive proceeding. I knew my position in relation to Messrs. Overend, Gurney, and Co. perfectly well. The least hesitation on my part, with regard to what they wanted me to do, might ruin me. Mr. Edwards was their factotum, and of course knew their designs. I cannot tell now whether he was really acting according to their instructions, or whether he was only serving Mr. George Lascaridi, with whom, as I have said, he had other transactions.

"Look here, Mr. Edwards," I said at last. "I repeat that nothing will give me a greater pleasure than for Mr. George Lascaridi to reap a benefit from this business. I told you he is my benefactor. But I should not be justified, either for my sake or his, in entering into any partnership with him until matters are squared up with his firms. Look here, what Messrs. Lascaridi and Co. claim from me:—

[*Translation.*]

London, 9th March, 1860.

STEFANOS XENOS, ESQ.,
London.

Enclosed you will find our two accounts balanced up to 31st December last year, which presents a sum of £2017 12s. 4d. and £1327 10s. 7d.—viz., £3345 2s. 11d.—to your debit, and we beg of you to take in consideration the payment of this debt of yours. Also, you have not rendered us your account concerning the freights of Aleppo and Beyrout, on account of which you have paid us a sum of money, but we have to receive still a considerable amount. On the other part, we want the above accounts settled to balance our books, and we shall be very much obliged if you send it this week.

Yours truly,

LASCARIDI AND CO.

This was my reply to them:—

[*Translation.*]

(Copying Book, Folio 61.)

March 9th, 1860.

MESSRS. LASCARIDI AND CO.,

In answer to your letter of this day, I return you the accounts which were enclosed therein, as I do not recognize them. According to our books we have to receive, not to pay. This affair is to be decided by Mr. Edward Watkin Edwards, in accordance with the arbitration bond signed by your Mr. George Lascaridi and me. Mr. Edwards is now examining the various transactions between *you* and me, and by his decision we shall abide. Before he comes to a decision, we will not acknowledge any account—not the losses on the grain. I never entered into any grain speculations with you for joint account, therefore I cannot be affected by any losses that may have been incurred in such speculation. All that I did was to cede to you ten sailing vessels, chartered by me on the understanding that their cargoes were to be sold on the same day for 34s. per quarter, taking it for granted that you had bought grain through Mr. Elefteriades at 13s. a quarter. It was solely on these terms that I let you have my ten vessels, which I could have re-chartered on the same day at 2s. a quarter profit. Besides this loss of profit on freight which you owe me, you make me another loss, under the pretext that your agent, Mr. Elefteriades, has abused your confidence.

With regard to the steamships Aleppo and Beyrout, there are nothing but the suspense accounts—namely, freights unpaid, such as that of Salik Pasha, agent to the Sublime Porte, or those of persons who have failed, or reduced freights paid by merchants whose goods were lost by your captains. Any balance that there may be we have passed to your account, and will be decided upon by Mr. Edwards. I must, however, make this observation, that Messrs. Fachri, Lascaridi, and Co., of Constantinople, in whose hands I placed the transaction of my business there, in accordance with the wish of your Mr. George Lascaridi, now detain my deposits of coal worth £6000. Messrs. Sursook, of Beyrout, have kept moneys of mine; and Mr. Elefteriades, whom I made my agent, also at the request of your Mr. George—and to do so I had to take my business away from my friends—has detained £1000 of mine, on the plea that you had refused to accept his bills.

Without entering into further explanations, I return your accounts, as I will not acknowledge them. I am ready to abide by the decision of Mr. Edwards, and to pay you to the last farthing. But I am grieved to inform you that, according to our accounts, the Greek and Oriental Steam Navigation Company has to receive, and not to pay.

Yours obediently,

(Signed)   STEFANOS XENOS.

In this state of affairs, how can you wish me to enter into a partnership with George?"

"Xenos, you know that these accounts are in my hands, so I give you my promise that Messrs. Lascaridi and Co. will never trouble you. I shall make George give you a guarantee that you will not be troubled by his firm."

Mr. Edwards then called his confidential clerk, Mr. March, who was in the next room, and wrote a duplicate draft of a new agreement, with the alterations that I made. It is as follows:—

97

. [*Copy Memorandum.*]

THE GREEK AND ORIENTAL STEAM NAVIGATION COMPANY.

1st *August*, 1860.

Present: Mr. XENOS,
Mr. EDWARDS,
Mr. MARCH.

Mr. Xenos will agree that a Partnership Deed shall be prepared by Messrs. Crowder and Maynard between himself, G. P. Lascaridi, and Mr. Aristides Xenos—G. P. Lascaridi undertaking that no claim shall be made against the above company by the firm of Lascaridi and Co., and that the latter shall withdraw the stop that their house at Constantinople has placed upon the property of the above company.

Under these circumstances, the shares of the partnership to be $\frac{39}{84}$ to Mr. Xenos, $\frac{39}{84}$ to Mr. G. P. Lascaridi, and $\frac{6}{84}$ to Mr. Aristides Xenos.

Mr. Xenos to attend to the management of the fleet; Mr. Lascaridi to the books and to finance, and to attend at the offices of the company for three hours at the least daily.

No bills to be accepted or cheques drawn, or any new business created involving outlay beyond the necessary expense of the fleet, except under the direction of Mr. E. Edwards.

Two banking accounts to be kept—one for sums exceeding one thousand pounds, against which the signatures of both Mr. S. Xenos and Mr. G. P. Lascaridi shall be required; and one for the working account, against which each of those gentlemen shall sign for self and partners.

Mr. Aristides Xenos to have a salary of £500 a year, in addition to his share of the profits.

(Signed)   STEFANOS XENOS.

E. W. E.
C. J. M.

I signed the draft, giving my brother Aristides a small share as partner in the concern.

Fatal to me was the moment in which I signed that document. Lascaridi had worked the official assignee, who had worked David Ward Chapman, and Chapman, Henry Edmund Gurney; and the great man

H

having uttered the decree, I was compelled to submit. Any hesitation on my part would have been construed into rebellion, and no opportunity for explanation would have been afforded me, as the system at the "Corner House" was to give audiences of not more than three minutes' duration.

Mr. Edwards, in his statement before the Lord Chancellor, said: "I have no doubt now that I did receive, at some time or other, money from Mr. Lascaridi, acting for and on account of Mr. Xenos, as commission on a loan negotiated with Overend, Gurney, and Co. for that gentleman; but that had nothing whatever to do with the subject of the arbitration, and was long subsequent to it." Now, this is something which I hear for the first time. I can understand now why Mr. Edwards forced upon me this partnership, which afterwards entailed a loss of £100,000 on the house of Overend, Gurney, and Co., as we shall presently see. From what money of the Greek and Oriental Steam Navigation Company were these commissions paid? Were these commissions paid before or after our partnership? In what books of the company, and in what form, were they inserted? Such voluntary revelations to me now are astounding.

Within a few days the deed of partnership was prepared and signed. From that date Mr. George Lascaridi became a fixture in my office. He superintended the cash department, and managed the finances, according to the wish of Messrs. Overend, Gurney, and Co.

Lascaridi had very quickly comprehended the relation in which I stood to Messrs. Overend, Gurney, and Co., and had consequently set to work, with the official assignee, to obtain, through them, a partnership in my business. When he found that all my debts were concentrated in oné place, and that my creditors were amongst the greatest capitalists of Europe, he saw that they, having advanced me £200,000 on ships and grain, would, for their own sakes, continue to support me. He did not misunderstand his interests. Independent of the £1000 a year that he had to receive, according to our partnership deed, from the Greek and Oriental Steam Navigation Company, this partnership would also serve him to give stability to his many worthless enterprises.

The day after that on which I signed the draft I received from Mr. Lascaridi the following letter:—

[*Translation.*]

August 2nd, 1860.

DEAR XENOS,

I have received your letter of yesterday late to-day, and I replied to it immediately, but before sending it I saw Mr. Edwards, who told me what you and he had agreed on yesterday together, which I find right with the exception of a few additions which he approves of himself. He told me not to send my reply, as you expressed regret for your manner to me last Saturday. In consequence, I send the past to oblivion, and on Monday next, on my eturn from Manchester, I shall come to your office and have a talk with you.

Yours,

(Signed) G. P. LASCARIDI.

CHAPTER XVII.

THE "BRITISH STAR."

THE reception that my two literary and political works, viz., "The Exhibition" and "Coining," had met in the East, and the favourable impression that their illustrations had made on the public mind, first suggested to me the idea of establishing, in London, an illustrated Greek newspaper. It should be literary, political, and commercial. With such an organ, placed, as I believed, beyond the reach of arbitrary power, I hoped to propagate throughout the East a knowledge of English institutions; I hoped to make those remote peoples, in whom I took so deep an interest, acquainted with the civilization of Western Europe. I believed that such information, diffused throughout those regions, would eventually replace Russian by English influence in the East. I would expose the disgraceful conduct of the Bavarian Court at Athens. I even ventured to hope that a voice travelling from the free soil of Britain might reach the ears of King Otho, and induce him to alter a policy that was virtually annulling the constitution he had sworn to defend. Ultimately the *British Star* was destined to become the chief pillar of the Greek and Oriental Steam Navigation Company in several respects.

I lost no time; I brought from Athens ten gentlemen whom I intended to employ as translators; I also brought Greek printers, and, of course, Greek type. Things were soon ready, and early in 1860 I established in London a Greek printing-office, and commenced to publish the *British Star*.

As the subscription to the *British Star*—£3 3s. per annum—was to be paid in advance, the Greek and Oriental Steam Navigation Company issued a circular to the public in the East, announcing the guarantee of its duration for two years at least. (See Appendix, No. 7.) I was not disappointed in my expectations.

This luminary shone for some time like a star of the first magnitude above the horizon of the East, but was suddenly obscured, and at length finally eclipsed, by a dense political fog with which Earl Russell contrived to fill the British Temple of Liberty.

How such events came to pass I shall narrate in due order. Let me now return to the fleet of the Greek and Oriental Steam Navigation Company.

It was about this time that the *British Star* began to make a great stir in the principal towns of the Levant. The monstrous errors—to call them by no worse name—of King Otho's administration were pointed out in its columns, and British institutions loudly advocated. Translations of long extracts from English newspapers and magazines, with miscellaneous information upon almost every subject, together with beautiful illustrations, made the *British Star* the veritable leader of the Greek press. That journal was so identified in the minds of the Greek people with the Greek and Oriental Steam Navigation Company, that they came to be regarded as national twins, and it must be confessed that the newspaper was no small support to the company, gathering all the shippers round it.

No sooner had George Lascaridi become my partner than he was suddenly seized with the itch for writing.

He would fain write the leading articles for the *British*,
Star. He had never written a line for the press, but he
did not understand why that should be an objection to
his writing a leading article. Lascaridi was what is
called a "clever fellow," and dealt out his opinions
freely on all subjects. But the manager of a newspaper knows how to discriminate between a voluble
talker and a thoughtful writer. I put an interdict
upon Lascaridi's attempts at "leaders." I also prohibited his interference with the administration of the
steamers. A serious collision between us would have
been inevitable had it not been that he, perplexed by
his other partnerships, came to a crash. But before
coming to a crash, he kindly tried to give me precedence, and to place me in that position before reaching
it himself.

He commenced operations by informing Messrs.
Overend, Gurney, and Co., through Mr. Edwards,
that the *British Star* would be the ruin of the Greek
and Oriental Steam Navigation Company—firstly, because the expenses of the newspaper were enormous;
secondly, because it would never pay; thirdly, because
its tone was so inflammatory that all the shippers
began to desert the company; and, finally, because the
Turkish, Greek, Austrian, and Russian Governments
were so hostile to me personally, that they would put
every impediment to the navigation of my steamers in
the waters of their respective kingdoms. He further
informed his employers that, since the finances had
come under his direction, he had discovered that the
affairs of the Greek and Oriental Steam Navigation
Company were in a state of insolvency. Nothing
would have been easier for me than to contradict
these statements had I known of them; but I did not.

and unfortunately those to whom they were made were not yet thoroughly acquainted with me.

I was suffering, though unconsciously, under the disadvantage of these slanders, when one of those epidemics that occasionally devastate the commercial world suddenly broke forth amongst the Greek merchants. Money became scarce, failures frequent, and terror general.

Messrs. Overend, Gurney, and Co., who about this time would have to renew my first bills, became alarmed. They asked Mr. Edwards to go through our books with Lascaridi and me, and report our position; and we met one night in the office of the Greek and Oriental Steam Navigation Company. Lascaridi was a little embarrassed. He was expected, though I did not know it at the time, to show the official assignee our insolvency. He began in general terms to deplore the bad state of affairs, but he offered no arithmetical proof of what he said. He then proceeded to charge the *British Star* and me with all kind of antipatriotic and uneconomical sins, and finished by saying that the Greek and Oriental Steam Navigation Company would soon be insolvent.

I was taken by surprise; I could not comprehend the motive for his conduct. My brother Aristides said to him, quietly: " Pray, sir, are you not a partner in this firm ? Had you been our most bitter enemy, you could not have behaved worse than you have done. If what you say is true, why did you not in the first instance make these communications to your partners? As a partner of the firm, you were bound in honour to do so. We might then have consulted as to what was best to be done."

Lascaridi looked confused. My eyes were opened;

I saw clearly how things were. I drew the books before me, and began to explain to Mr. Edwards the position of our affairs. Lascaridi, to support his slanders, endeavoured to depreciate the value of the steamers. I spoke to him, before Mr. Edwards, in the strongest terms. Mr. Edwards, as an accountant, not only understood figures perfectly well, but was prepared to save the business if he saw there was no systematized deceit. My explanation fully satisfied him. I then turned to George and said:

"George, I have only lately learned the true state of your private affairs, but I could never have believed you capable of acting as you have done. I now leave Mr. Edwards and his employers to find out the wrong they did me by forcing this partnership. They will shortly know all about you."

CHAPTER XVIII.

THE PENELOPE.

WE all four left the office together. It was the 28th January, 1861, a very wet night, and, whilst we were waiting for a cab, I observed that Lascaridi was trying to separate from us, and to go on with the official assignee. I said to myself: "If he succeed in doing this, he will perhaps cause him to change his mind again before morning;" so I determined that they should not leave together. A cab happened to pass at the moment. I stopped it, got in, and invited Mr. Edwards to take a seat. He stepped in, and asked Lascaridi to join us; but I interposed, and said that his road lay in a different direction, and at the same time ordered the cabman to drive on.

"What motive could Lascaridi have had for behaving to us in this way?" I asked of Mr. Edwards, soon after we had started. He remained silent. "And, supposing there were any truth in his statement, surely he ought to have communicated his discovery to us, who are his partners, so that we might consult together as to what was best to be done; but to go first to Chapman or you without our knowledge—I cannot understand it. Perhaps he has some other object in view."

"Never mind! you peppered him enough this evening; forget it now. Where are you going to-night?"

"I am going to get some dinner; I have eaten

nothing since eight o'clock this morning. It is killing work to have to look after twenty steamers, and to battle against such intrigues as I have had against me since I first made the acquaintance of the Gurneys."

"We all kill ourselves for money," remarked the official assignee. "I was ill in bed three days ago, when the Overends dragged me out of my bed about your business. Will you come and have a little dinner with me at the Garrick?"

"With pleasure," I answered.

"We must, however," said he, "go first to Victoria Street, to leave some papers, and wash." He consequently ordered the cabman to drive to No. 27, Victoria Street Flats. It struck me as a strange coincidence that he should occupy the first flat, where I had myself been living in 1857, just three years before.

"This is my town residence," said the official assignee. "It is very small. We are located permanently at Brighton, where, if you come to see us, you will find things in a different style."

"David Ward Chapman is living there also, is he not?" said I. I was washing my hands in my old bed-room.

"Yes," answered he; "Mrs. Edwards and Mrs. Chapman are great friends. Chapman lives in great style at Prince's Gate, Hyde Park, where he resides when in London. He keeps ten horses. He has his house always full of visitors, and gives the most delicious dinners. He is the most splendid fellow I know. He lives at the rate of from £15,000 to £20,000 a year."

"Twenty thousand pounds a year!" exclaimed I.

"If all the five partners live at that rate, what with office expenses, the profits of the 'Corner House' must be over £120,000 a year."

"You forget the extent of their business," said Mr. Edwards, "and the amount of their capital. The Gurneys must be worth at least ten millions sterling."

"Fabulous!" I replied. "Are the Chapmans Quakers?"

"Cross breed," said the official assignee.

"Are you a Quaker, Mr. Edwards!" I asked.

"Not altogether," replied he; "but I have a great deal to do with Friends."

We then left Victoria Street for the Garrick. It was past eight when we arrived at the club, and we found it nearly deserted. Mr. Edwards ordered dinner, which, when it came on the table, I attacked like a half-famished man, as I had been too busy since early morning to find time to taste anything. I also did honour to the splendid hock of the Garrick.

"I cannot give you any champagne or red wine," said Mr. Edwards, "as the doctors have ordered me to take nothing but hock."

"I am content with the hock," said I, although I thought it rather selfish of Mr. Edwards not to offer red wine to his guest because he had been forbidden to drink it himself. After dinner we retired to the smoking-room, where we were quite alone. The conversation turned on the Greek and Oriental Steam Navigation Company and Messrs. Overend, Gurney, and Co., but more particularly on the *British Star*. I told Mr. Edwards confidentially that the policy of the paper was not selfish, as Mr. Lascaridi had stated—it was to oust, if possible, the then existing Bavarian dynasty, and to offer the throne to Prince Alfred. I also told him that one or two Englishmen of high position encouraged me in my project, and that, if I attained the object I had in view, I should render immense service to my country, and consequently procure certain privileges for the Greek and Oriental Steam Navigation Company, which would bring it great wealth. Mr. Edwards began to understand my policy, and seemed highly pleased. He then expressed a wish to make a cruise in the Grecian Archipelago. I told him the scenery there was fairy-like, and would well repay a visit. At that time I had two yachts, one a sailing and the other a steam yacht; but I am so bad a sailor, and had so little time, that I never used them. They were quite new, and lying in Victoria Docks; they had never yet been to sea. I mentioned this to Mr. Edwards, who said, "Will you lend me the steam yacht for a short cruise?"

"All right," said I; "you can have her altogether, as she is of no use to me."

"Thank you, you are very kind," said the official assignee, shaking me by the hand.

We had no further conversation on the subject. After a short pause Mr. Edwards said, "Xenos, do you know what is my mission?"

"Not I."

"To become a very rich man; I see that on every side I make money, and that from every side money comes to me."

"You are a most fortunate man," said I.

"It is not that, it is mathematics, my friend; my talent is finance. The Messrs. Gurney wished me to go into Parliament, but I have declined for the present, and until I have made a large sum of money."

At about half-past ten we left the club and walked

towards Piccadilly. I wished to go home, as I was tired and exhausted, so I took a cab and was driven to Kensington.

The next day I received a short note from Mr. Edwards, requesting me to send him the papers of the yacht Penelope, of which I had been so kind as to make him a present. I had forgotten all about the matter, and this note made me think a great deal. I requested Mr. Edwards's clerk to remain in the outer room a little while, and sent for my brother Aristides. When I told him what had taken place, he advised me not to send the papers. "If you make him a present of £2000 in that way," said my brother, "he will think that either you are a fool, or that you give him the yacht as a bribe."

"I look at the matter in a different light," said I. "I know no Overends in this matter. He and Chapman are everything; they have all the power. If I do not send the papers of the yacht to Edwards, he will tell me that I am a man who does not know how to keep his word. He will think it a breach of faith. All my fleet is now in his hands and power. The refusal to give him the yacht might make him spiteful, and cause him to ruin us, particularly now that Lascaridi has declared himself an open enemy."

"I would not give her," replied my brother, "were I in your position."

I reflected a short time longer, and then, much against my will, sent him the yacht's papers. It is needless to add here that Mr. Edwards kept the yacht at my expense for two months after he had become her owner. He registered her only on the 28th of March, 1861.

One day my friend Mr. G. B. Carr, on entering

my office, handed me, with a smile, a short paragraph, which he had cut from some paper, saying at the same time, "There is a little news for you." I took the paragraph, and read as follows:—

Mr. E. Watkin Edwards, accompanied by his friends, Mr. Harrison Ainsworth and Mr. Lyster O'Beirne, arrived at Paris on Saturday last, in his steam yacht Penelope, belonging to the Royal Victoria Yacht Club, Ryde, having successfully navigated the Seine. This, it is believed, is the first English yacht that has ever been moored off the Tuileries, and is, therefore, an object of great curiosity to the Parisians.

"I always suffer dreadfully at sea," I remarked; "but he might, at all events, have sent me an invitation, particularly as he was sure I could not have accepted it."

"Not he," said the kind old man, with an expression on his face which I shall never forget.

How little I knew my man when I gave him the yacht in 1861 the following letter will show. When the fall of the Anglo-Greek Steam Navigation Company came on me quite unexpectedly, I was pressed on every side for money. Being almost at my wit's end, I said to the ex-secretary of the defunct company, "I have a mind to write a letter to a gentleman to whom, in my days of prosperity, I made a present of a steam yacht, some hundreds of pounds, and an Arab horse, to ask him to help me. It is said that he is immensely rich, unless he has lost his money in the panic, or through the failure of Messrs. Overend, Gurney, and Co."

"Then surely he will do it," said my friend.

"Not he," I answered, using Mr. Carr's phrase. "I wish to have you as a witness to his ingratitude. Write the letter yourself, and I will sign it." The letter was written and signed. The following is a copy, word for word:—

9, *Fenchurch Street, June* 18, 1866.

E. W. EDWARDS, ESQ.

DEAR SIR,

I have had lately to go through many difficulties, as many others in the City of London. The London and Westminster Bank finding the opportunity to sell my estate (Petersham Lodge) on account of £500 accumulated expenses, after I arranged the transfer to the North British and Mercantile Insurance Office, can you lend me that sum on additional security, and save the estate? The bank advance is £5500. The estate cost me £7000 as copyhold, £1000 to franchise it, and I have spent £3000 in new buildings, and since I bought it property has increased in value 25 per cent., so there is ample security. I want only the £500 for one year, till the affairs of the Anglo-Greek Steam Company are settled, when I expect to have a large sum of money that is due to me. An immediate reply will oblige me.

I am, dear Sir,

Yours truly,

STEFANOS XENOS.

Mr. Theophilus Kokos, then my clerk, who was present at the above conversation, took the letter to Mr. Edwards's commercial offices, 81, King William Street. In a short time he returned with the answer that Mr. Edwards was not expected there that day. Mr. Kokos added, however, that the clerk to whom he handed the letter took it into the inner office, remained there some time, and then came back without it, so that consequently he believed that Mr. Edwards was in. On hearing this, I said I would go myself. When I arrived the door of the private office was half open, so that I could hear Mr. Edwards's voice within; the clerk could not, therefore, deny him to me. I sent in my name, and the reply was that Mr. Edwards hoped I would excuse him, as he was too busy to see me that day. I then wrote on a slip of paper "£250 will do instead of £500." The clerk took it in, and returned saying "No answer."

Reader, was the yacht a friendly gift, as Mr. Edwards asserted before the Lord Mayor at the Guildhall, and before Mr. Commissioner Holroyd at the Bankruptcy Court? If so, Mr. Edwards was guilty of that ingratitude for which the Persian law-makers decreed decapitation as punishment. If it was not a gift, but that Mr. Edwards thought he had a right to it, was it the realization of the dove and the hawk?

The Penelope was 40 tons gross measurement, $79\frac{1}{2}$ feet long, $12\frac{9}{10}$ broad, $5\frac{7}{10}$ deep, and 25 horse-power. She was not given to me as a present, as asserted by Mr. Edwards in his statement before the Lord Chancellor. I contracted for her, as well as for another smaller sailing yacht, the VIEROUKA, with Messrs. A. Leslie and Co., of Gateshead-on-Tyne, and paid that firm the sum of £2360 for them. The Penelope had cost me, including stores, insurance, wages, and other charges, about £1800 when I gave her to Mr. Edwards. I never promised her to Mr. Edwards during her construction, as he asserted in his statement, or on any former occasion. I promised her to him for the first time on the evening that we dined together at the Garrick Club, and that was the 28th of January, 1861, up to which time, from November, 1860, she had been lying idle in the Victoria Docks, as can be proved by reference to the books in the dock offices, and the letter of her builder, Mr. A. Leslie. (See Appendix, No. 48.)

CHAPTER XIX.

DEEPER IN THE WATER.

The commercial crisis, or the Greek commercial epidemic, spread so fast in 1860, that each morning brought intelligence of new victims.

That crisis had a political origin; perhaps the time is not far distant when it will bring forth a political result.

That epidemic had its rise in the seraglio of the Sultan, and in the courts of the high Pachas.

Before the Crimean War, Turkey had no foreign debt. During the Crimean War she was in want of money, but the arrogant Turk would rather do anything than borrow from the Franks. In his difficulty, he applied to the Greeks of Constantinople. He would give any rate of interest—which, in plain English, means that the principal would never be paid. About a dozen Greek firms of Constantinople, having houses in London, undertook the job. It was executed in the coolest manner. One firm at Constantinople drew on its London house, and passed the bills to the other firm, who gave its own paper in exchange. The bills were sent to Lombard Street, melted, found their way back to Constantinople, and were poured into the Imperial Treasury in a stream of liquid gold. The Turk paid on these transactions from 20 to 50 per cent. The loans were renewed, with additional interest, and the bills met by similar accommodation following quick one on the other; and if, within

I

twenty-four months, the merchant at Constantinople succeeded in making the Sultan pay, out of his revenues, the interest alone, all was right.

To Greek merchants who had large credit this species of commerce appeared not only very profitable, but involving no risk. It was more lucrative and easier to manage than the grain trade and the trade in colonials, in which they had been brought up. But the business was at length overdone; too many had engaged in it, and a crisis was the consequence. The small scraffs at Constantinople had commenced to imitate the great merchants, and Lombard Street was inundated with Greek paper. The bankers began to see through the game, and became apprehensive of danger. Messrs. Overend, Gurney, and Co., as may be supposed, held no small share of this paper. Now the *British Star*, as the exponent of Greek interests, had openly denounced the Greeks engaged in this paper commerce as the destroyers of Greek credit and Greek trade; it is easily understood, then, how some of the Greeks of London formed a formidable clique, and tried their utmost to injure me in the opinion of Messrs. Overend, Gurney, and Co.

The Greek public, ignorant of the secret springs that worked this machinery, bought the bills of these houses in the different exchanges of the Levant, and remitted them to the smaller Greek houses of Western Europe. At length the difficulty arose. What was to be done with unnegotiable paper? And so the epidemic seized Greeks in London, Manchester, and Liverpool, who were wholly guiltless of the original sin. It was a terrible affair. Many of the originators of that catastrophe died, commercially, miserably; others are to this day creditors to the honest Turk for large sums; but some few—*les plus habiles*—contrived to become the indispensable favourites of the grave Mussulman. These gentlemen assisted the Turk to consolidate all his trifling internal debts into a stock importable and marketable in London. The success of the undertaking so astonished the Grand Vizier, that he one day asked Mr. S—— why it was that Englishmen paid in London, for these *caïmes*, three times what they were worth at Constantinople.

"Your Highness," replied the flatterer, "Englishmen wish to prove that their love for Turkey is a real, not a platonic sentiment."

"Then I foresee that they will one day monopolize her," said, with a shrewd smile, the Europeanized Mussulman.

About this time the affairs of the Greek and Oriental fell again into an alarming condition. There were several causes for this. In the first place, the Levant export trade had come to a standstill, and the import trade was seriously paralyzed; and secondly, and the most important, Lascaridi, in financing our company, had discounted or exchanged about £12,000 of our good acceptances with those of some Manchester, Gibraltar, and London houses with which he was connected, and which failed a few days after the transaction.

"The larger the ship the greater her share of the storm," says the Greek proverb; and such the Overends felt to be the case in that worse than equinoctial hurricane. The winds blew from every point of the compass, lashing the commercial sea into fury. A hail of bills fell simultaneously from north, south, east, and west, but the missives were quickly melted on the deck of the gallant ship. A profound darkness enveloped the horizon; the atmosphere was scarcely respirable; the thunder rolled moodily overhead; and the forked lightning served to show the horrors of the gloom. Sinister noises were heard on every side. The brave commanders, Henry Edmund Gurney and Robert Birkbeck, resolutely faced the storm, and tried to reduce to form the chaos in which they found themselves. As for David Ward Chapman, he could scarcely keep on his legs. He was in a fainting condition when he heard the creaking of the straining timbers of the vessels of the Galway Company, and those of the Greek and Oriental, of Zachariah Pearson, of the East India and London Steamship Company, and others congregated round Messrs. Overend, Gurney, and Co., and heard their crews calling for help. Poor David, bewildered and confounded, thought the end of the world was come. He was bowed to the earth in penitence for his past faults. He blamed his favourites, and he blamed himself; though, sooth to say, at the first appearance of fair weather he freely forgave both parties.

Happy Arthur Chapman! Retired within the luxurious saloon of the ship, he knew nothing of the storm that raged above. Nay, he refused to believe aught of the hurricane until the victims of the tempest and the floating wrecks were pointed out to him. And when he had realized these evidences, he would merely lisp, in his arrogant manner, that the victims were "rogues or fools."

During this fearful state of things, Mr. Edwards, the *pontifex maximus* of Basinghall Street, could find no leisure for my affairs. He was busy from early morning to latest eve in administering the last consolations to the victims of the storm, or else helping them to the deck of the great ship, running thereby the risk of overloading her at a moment when she needed to be lightened. I complained to Mr. Edwards of Lascaridi's reckless mode of financing, and declared that I would not recognize the acts of this partner and financer that had been thrust upon me.

A rather sharp contest took place one day between Mr. David Ward Chapman and me. It occurred in this way. I had begun to import grain very largely from the Danube on our own account. The bills of lading for several thousand quarters of this grain were expected to arrive daily in London. I had some heavy payments to meet, and as Mr. Lascaridi, my second financer, had by this time begun to come but very rarely to the office, I was obliged to go in person to the "Corner House" to ask for funds. When I arrived there were several Greeks there, and when Mr. David Ward Chapman saw me, he said, in an alarmed manner, "Walk in there." After keeping me waiting for some time he entered, and, before giving me time to explain the cause of my visit, cried out, "You are come again for money. Are you not satisfied with what you have already had from this office? You are for the Old Bailey."

"Are you mad?" said I, at the same time preventing him from leaving the room. "What right have you to speak to me so? It is your house that is systematically breaking me down by your enormous interests, and by the acts of those whom you have appointed to finance my business. I, who have a right to complain, say nothing, and you scream out. Have you not ample security in ships, freights, and grain, for the money which you have advanced me? Perhaps you prefer to be the holders of hundreds of thousands of unsecured bills, such as you have held during this crisis. What will you get out of those who are failing daily?"

Mr. Chapman cooled down, and left me; he returned, however, in a short time, in company with Mr. Edwards, to see what I wanted. I explained my wants to them, and at the same time showed them the worthlessness of Mr. Lascaridi's accommodation bills, which were sure to bring a heavy loss to the Greek and Oriental Steam Navigation Company. I added that I should not recognize the bills, as Messrs. Overend, Gurney, and Co. had forced me to take Mr. Lascaridi as a partner. Whilst I was speaking Edwards and Chapman kept looking at each other in silence. "I was the happiest of men," I continued, "when I was a ship-broker; I had a very profitable business, and no responsibility. Now, besides the tremendous labour of working this great fleet profitably, I am also called on to make superhuman efforts to prevent the ruinous consequences of some of the acts of those appointed by the firm to guard their interests. I bide my time, for the day will come when you will learn that you have devils in human form connected with this office."

"To whom do you allude?" said Mr. Chapman, looking sharply at Edwards.

"I do not know," I answered.

Mr. Henry Edmund Gurney then entered the room, and spoke in a low tone to Mr. Chapman about some other business, after which they left the room together, leaving me alone with the official assignee to arrange my matter.

CHAPTER XX.

MY THIRD FINANCER.

Mr. Edwards, the emperor of the "Corner House," now changed his minister of finance. Seeing that Lascaridi could not attend to our business, he appointed Mr. Henry Barker to superintend our finances. Mr. Barker was to present at Lombard Street, every month, a list of the payments of the Greek and Oriental Steam Navigation Company, and he was to receive a cheque for the amount. Mr. Henry Barker's establishment was, as I have observed, started by Mr. George Lascaridi—a fact of which I was at that time ignorant, as everybody else. This new appointment was a fortunate event for Mr. Henry Barker. He was just then on the verge of bankruptcy; his estate was in the hands of Messrs. Overend, Gurney, and Co., and by adroit management of the funds that now began to pass through his hands, he raised himself to the position in which we shall shortly find him. My payments during the course of the month amounted to from £20,000 to £30,000. Mr. Henry Barker having received a cheque from Messrs. Overend, Gurney, and Co., lodged the amount at his banker's. He went every day at four o'clock to my banker's—Mr. Marshall, of the Bank of London—where he had an understanding to keep until four o'clock all my crossed cheques and bills and pay them. In this way, Mr. Henry Barker was able to keep a splendid balance at his banker's, and raise his own credit, whilst he was sending that of the Greek and Oriental Steam Navigation Company to perdition. But this was not all. In return for the payments thus made by Messrs. Overend, Gurney, and Co., I was bound to give them the freights and earnings of my steamers. The orders were that I should hand the freights to Mr. Henry Barker, who would give Messrs. Overend, Gurney, and Co. a single cheque, comprising all details, so that it should seem as though the transactions were exclusively between these parties, my name not appearing at all in the books of Messrs. Overend, Gurney, and Co. I do not know who organized this system, but I believe it was adopted by the great money-lenders with others to whom they made advances. Mr. Henry Barker licked up a little honey out of these transactions. He contrived, during several years, to keep in his hands about £12,000, concerning which there were disputes and disputes year after year. I am ignorant even now how he settled with Messrs. Overend, Gurney, and Co., in 1865.

You, reader, can easily understand my indignation. The system pursued towards me was tyrannical and destructive. Whilst Messrs. Overend, Gurney, and Co. held my property as security for certain loans, the money they advanced was paid over to a third party —a gentleman of very small means. But, worse than all, they had forced me to sign the following letter, drafted by their head clerk, Mr. Bois, holding them harmless in whatever way Mr. Henry Barker might apply the money advanced:—

London, 19 January, 1861.

MESSRS. OVEREND, GURNEY, AND CO.,

In consideration of your agreeing to advance to us various sums, as they may be required, to meet our engagements, we hereby authorize you to hold all the securities deposited with you as collateral security for the same, and engage to further assign to you the SS. Palikari and Odyseus Androutzos, when in my possession, and do further authorize and direct you to pay the said various sums to Mr. Henry John Barker as he may call for them, and have instructed the said Mr. Henry John Barker to sign and give acknowledgments for the said sums, and hereby agree that his signature to the said memoranda of advances shall make them as valid as if they were signed by ourselves.

We are, Gentlemen,
Yours very truly,
(Signed)   STEFANOS XENOS.

I was exasperated at this, and called at the "Corner House" to put an end to this disgraceful conduct of Edwards's financings. About this time I began to suspect that Henry Barker and George Lascaridi were connected closely together in business. It was no easy matter to get a private interview with any of the partners at a moment when the storm was blowing so violently. On making my appearance, Mr. David Ward Chapman, who thought that I was going to ask for fresh loans, or to announce the sinking of my fleet, was again seized with terror. He called out hastily: "Walk in there—walk in there," pointing to the dark small room. Being a dull day, that room was more than usually gloomy. Often and often had I sat there for hours, conning the *Times*, whilst I waited an interview with one of the mighty money men. On the present occasion I wished for a short *tête-à-tête* with Mr. Henry Edmund Gurney, but I soon found that that wish could not be gratified until the commercial storm should have subsided. I left in despair.

As may be expected, that storm did not subside without producing serious results for those who, whilst it blew, found themselves unprepared outside the port.

It was about this time that George Lascaridi, one' very frosty morning, entered my office, and in a lively, agreeable manner informed me that he intended to suspend payment. He further added that he had already given notice to his bankers not to honour his acceptances.

"Stop payment!" I explained; "what payment? What is the amount of your private debts?"

"Only £15,000."

"You surely do not mean to stop payment for the sake of £15,000. But, George, when you entered into partnership with me, you told me, on my asking, that you had no debts. How can you, in the course of a few months, have contracted these debts? George, if you stop payment, you will ruin this company. Your creditors will fall on my poor steamers, because you are now my partner."

"I can't help it. Good-bye, good-bye. This is the advice of Mr. Cotterel, my solicitor."

And with quite a holiday air he walked lightly out of the office. I was petrified. After a minute I seized my hat, and pursued George. I lost him in Fenchurch Street. I hurried to Mr. Henry Barker's office. He was not there. Up to half-past three I sought him in vain. I was in agonies. I thought all was over. I at length decided upon going to Messrs. Overend, Gurney, and Co., and acquainting them with what had occurred. Turning the corner of a street, I met Mr. Henry Barker.

"Have you heard what our friend George intends to do?" he asked, with an air of great satisfaction.

"Then you know everything?" I remarked.

"Oh, yes!" replied Mr. Barker. "He called on me this morning, and showed me the letter he wrote to his bankers. Assuredly," he added, "George is a most mysterious man. At first I did not know what to do; but, reflecting on the great interest Messrs. Overend, Gurney, and Co. and the Greek and Oriental Steam Navigation Company have in his affairs, I went to Chapman, and told him all."

"Well?"

"Of course David Chapman was awfully alarmed. He begged me to go through fire and water to find and bring him to him."

"Did you find him?"

"Yes; fortunately I found him at my office. I took him to Lombard Street. Chapman, looking quite alarmed, said: 'Mr. Lascaridi, what is this you intend to do?' 'I must suspend payment,' replied George. 'My assets are larger than my debts, but I cannot realize them at this moment; so my solicitor advises me to suspend payment. My debts are about £18,000. I cannot any further involve the house of Lascaridi and Co.'"

The substance of Mr. Barker's story was this— Mr. David Ward Chapman, alarmed at Lascaridi's announced intention of suspending payment, offered him a cheque for £20,000 to pay his debts, which he refused, declaring that he would not take Chapman's money. After many refusals on the part of Lascaridi, Mr. Barker stepped forward, and quieted his scruples about accepting the money by proposing that he should make over his assets to Messrs. Overend, Gurney, and Co., who in return should become responsible for his debts.

I understood the game. Messrs. Overend, Gurney, and Co. held Lascaridi and Co.'s acceptances and indorsements to the amount of £300,000 sterling.

Mr. George Lascaridi's suspending payment would be fatal to the old firm, the principal partner of which, learning his doings, had come to England, and obliged him to resign. He had caught Messrs. Overend, Gurney, and Co. in his net. They were obliged to stand by him.

"We sent for Mr. Edwards," continued my third financer; "George and I are to meet at his house to-night, to arrange the affair. But I must leave you. I am going to George's bankers, to lodge money to cover this day's payments. Good-bye. Come to Edwards's to night, if you like."

I felt relieved. The weight that had pressed so heavily on my mind all day was removed. However, I was far from foreseeing the *dénouement* of this tragico-comedia.

CHAPTER XXI.

HOW EIGHTY THOUSAND POUNDS WERE LOST.

IT was, I remember, a cold frosty night, the streets covered with snow, when, about eight o'clock, I left my house to go to Victoria Street, Pimlico, where Mr. Edwards still lived. On my arrival I found the triumviri smoking the best Havannahs, and sipping wines of the finest bouquet, a variety of which stood on the table. It was a pleasant mode of settling the accounts of an insolvent gentleman. Lascaridi exhibited no emotion. He was fortified with a large dose of philosophy. The old Athenians used to say: Πωγωνοτροφία φιλόσοφον οὐ ποιεῖ (to grow a long beard is not to become a philosopher); but Lascaridi, with his fine beard, looked an exception to the rule, as he sat calmly giving the details of his debts to the representative of the great *depositari*. The calculation being made, it was found that his indebtedness, instead of being £18,000, amounted to £40,000. This revelation fell like a thunderbolt on Edwards and me. Edwards sat motionless, as if suddenly petrified, pen in hand and cigar in mouth, his eyes fixed on the figures inscribed on the paper before him. I was standing before the fire, but felt no warmth from the position—the live coal seemed to throw off no caloric. Barker, holding a glass of wine between his finger and thumb, sat with his eyes modestly cast on the floor. Lascaridi showed no emotion, except what might be detected in the sidelong, furtive glance which he occasionally cast around, and a slight expression of contempt about the lower part of his face. Otherwise, he was calm and dignified. He looked the hero of the drama; but his heroism was without wisdom, as his contempt was devoid of power.

Our profound silence was shortly interrupted by three small voices down in Victoria Street. Of course it was not a *barcarolla* of the gondoliers of Venice, nor the *canzonette* of Don Giovanni. I looked out of the window—it was a splendid moonlight night—and saw a poor woman walking with two barefooted children in the snow. Her voice was sharp but tremulous. Her wretched appearance, and the feeble treble of her shivering children, contrasted painfully with the sentimental tenderness of the ballad she fain would sing. At the moment that this miserable group passed, Mr. Edwards decided upon giving Lascaridi the trifling aid of £40,000. Strange contradictions in the human heart! Not one of us thought of helping the poor woman and her children, to whom a glass of wine from our table would have been as the elixir of life on that cold night.

Mr. Edwards now informed us that the alarm of Chapman, on hearing that George was going to stop payment, was principally caused through his anxiety about the Greek and Oriental Steam Navigation Company.

But to continue the tale of Lascaridi's affairs. I soon saw that if he underrated his debts he overrated his assets. According to his calculations, the manufactured goods that he had at Gibraltar were alone sufficient to cover what he owed. Then Mr. Barker proposed that he and some of Lascaridi's relatives should each become guarantee for a limited sum.

Lascaridi's business being thus comfortably arranged, Mr. Edwards asked him what he intended to do, now that he must retire from the Greek and Oriental Steam Navigation Company, and was no longer a partner in the firm of Lascaridi and Co.

"Well, I intend to ask Messrs. Overend, Gurney, and Co. to give me the management of one of their financial establishments. You know my capabilities in financial matters."

"They are not likely to do that," observed Mr. Edwards, grave as a judge.

"But I understand finance so well," said Lascaridi.

Was I in the presence of the Theban sphinx? The riddle of Lascaridi's private affairs was gradually unfolded. Within two months I learned that his debts, instead of £40,000, amounted to £100,000. The Overends paid all, and so the riddle was solved. I believe that the £80,000 which figures in the accounts of Messrs. Overend, Gurney, and Co., Limited, as the debt of Manuel and Co., is this very amount. Mr. Alexander Basilios Manuel was a clerk, and distant relative of Mr. George Lascaridi. About the year 1858 or 1859 he was sent by Mr. Lascaridi to Gibraltar and Tunis; when he arrived there Mr. Lascaridi consigned large quantities of Manchester goods to him. These goods were taken by Messrs. Overend, Gurney, and Co. as security, on undertaking to liquidate the private estate of Mr. Lascaridi. Mr. Edwards, in arranging this matter, never examined these goods to see what they were worth—never asked to see a list of them, nor their invoices, nor by what ships they were sent out, nor whether they were marketable, nor even whether they would fetch a price in any way approximating the sum of £40,000 for which they were given as security. Not even did he trouble to inquire whether £40,000 was the full amount of what was due from Mr. Lascaridi to his creditors. No written agreement was drawn up; the magic "All right" was all that fell from the lips of the great financer of Basinghall Street. I never recollect so bad a transaction as this on the part of the official assignee, who had been especially appointed to watch over the interests of Messrs Overend, Gurney, and Co. Poor Gurneys! they were utterly ignorant of what was being done, and what is worse, no one who had their interests really at heart could get a chance of speaking to them. Besides, Edwards was so powerful that, if any one had dared to do so, it would only have brought about his own destruction.

I have said that I thoroughly believe the debt of £80,000 placed against the estate of Manuel and Co. was really that of Mr. Lascaridi, for, after the Victoria Street scene, Mr. Edwards appointed some Englishmen with Manuel to superintend the Gibraltar business. Many of these goods, being unsaleable at Tunis, were transhipped to other places at enormous expense. The result was, that Mr. Edwards caused the "Corner House" to embark in the business carried on by the Manchester Greek houses—*i. e.*, the export of Manchester soft goods to the Mediterranean. The loss of this £80,000 was due entirely to Messrs. Edwards and Chapman—who were, in fact, in a greater panic about Mr. Lascaridi stopping payment than he was himself—and was undoubtedly the result of the partnership that had been forced upon me on the 1st of August, 1860.

Messrs. Overend, Gurney, and Co. discovered that

they held, and yet did not hold, the acceptances of Lascaridi and Co. for several large sums. I saw, too, why Lascaridi had been so anxious a few weeks before to break down the *British Star* and the Greek and Oriental Steam Navigation Company. Apprehending that, in consequence of his private debts, he should be obliged to suspend payment, he hoped, by involving me in his ruin, to fall with the *éclat* of martyrdom in the national cause. When he saw that I had convinced Mr. Edwards of the solvency of my concern, then he determined to suspend payment. His design was, like another Samson, to lay hold of the two pillars of the temple, and bury me and his partners in the *débris*. Happily for him, his good angel Henry Barker stepped in, and, for his own sake as well as Lascaridi's, induced Messrs. Overend, Gurney, and Co. to save him from bankruptcy by paying his debts.

Still Lascaridi was not satisfied.

When Messrs. Overend, Gurney, and Co. discovered that the acceptances they held bore not the veritable signature of Lascaridi and Co., but of G. P. Lascaridi and Co., they complained bitterly. Lascaridi retorted by saying that he had informed both Mr. Edwards and Mr. Barker that the acceptances were not those of the old firm, but of his new firm. He further threatened to give publicity to the affairs of the Galway Company. And all this at a moment when Messrs. Overend, Gurney, and Co. were paying the last claims of his creditors. Finally, and as the last act of this comedy, Lascaridi issued a circular, dated the 30th of March, 1861, announcing that he had taken up the business of insurance broker. Over the front of his office he painted in large letters, after the fashion of tradesmen, his name and new occupation. I do not remember, any merchant or broker in the City ever having made a similar epigram to announce his occupation. I never knew the object.

I leave any one to imagine what a busy time this was for the *mouches de cour*, the *renards de cour*, and all the courtiers that swarmed about the Lombard Street palace. Flatterers are the first to gossip about their patrons' shortcomings. Sinister rumours touching the solvency of Messrs. Overend, Gurney, and Co. were soon afloat; dépositors hurried to withdraw their deposits; but the great money men rose in their strength, and met the emergency.

CHAPTER XXII.

A BREACH OF FAITH.

THE day after the scene in Victoria Street, I called early on Mr. Edwards at his house, and told him that if the firm were going to pay the debts of Mr. George Lascaridi, to save him from bankruptcy, they must insist on a dissolution of partnership, by mutual consent, between him and me. " You remember the objections I raised when first you asked me to take Mr. Lascaridi as partner, and the scene which took place in my office about the *British Star*. You will now recognize the sense of my objection. This affair has cost me several thousand pounds; there is a loss of £4000 by Bello Brothers, for whom he discounted without my consent, and a loss of £8500 by the house of Vlagomeno, of Manchester. Both these houses have stopped payment. There is £1000 on S. C. Agelasto; he is going too. There are the bills of Caragianaki, Tzaliki, D. Rodocanachi, and others; all of these are going."

"You are right, Xenos; it was a mistake," said the official assignee. "I shall insist that he accompanies you and your brother to the office of Messrs. Vallance and Vallance, of Essex Street, Strand, and there dissolves the partnership by mutual consent."

He was in a hurry to get into the City, and gave me a seat in his brougham. We drove together to Lombard Street, where I left him. I went to my office. Mr. Lascaridi had not been there for several weeks, so that our partnership did not last quite four months.

At length the happy day came when Lascaridi, I, and my brother, met at the solicitors', Messrs. Vallance and Vallance, to cancel, by mutual consent, the deed of partnership. I felt as if a mountain weight were lifted from my breast. It is true that the property was burdened with £12,000, owing to Lascaridi's accommodation bills, with £12,000 on Mr. Barker's part, and with other large sums. Still I felt confident that, within a few years, I should be able to clear off these incumbrances, if left untrammelled. It was a pleasing prospect, and a powerful stimulus to action. But, alas! what says the proverb—" One kindles the fire, another is burnt in it."

Before the commercial storm had subsided, I one day saw the sale of a fleet advertised. The description of the steamers corresponded with mine. In the course of the day my broker, Mr. John Preston, called to say he had heard that Messrs. Overend, Gurney, and Co. were negotiating to sell my steamers to Mr. Zachariah Pearson. I could not credit the intelligence. My verbal agreement with Messrs. Overend, Gurney, and Co. was, that when the bill of £170,000 was due, it was to be renewed from time to time; at all events for one year. Nor was this an extraordinary accommodation, considering that I had paid £30,000 bonus, £3000 commissions, and 5 per cent. interest for the first six months. When the first term expired, Messrs. Lascaridi and Co.—viz., the other partners of Mr. George Lascaridi—were called upon to renew. They at first refused to accept again the responsibility which he had, without their knowledge, thrust upon them. They were, however, ultimately compelled to renew, no alternative remaining except to pay the bill for the enormous sum of £170,000. It was after this arrangement that Messrs. Overend, Gurney, and Co., being pressed for money, on account of the exceptional run of the depositors, broke faith with me, and, by the advice of the great Confucius, Edwards, tried to turn my steamers into money.

Any one connected in business with Messrs. Overend, Gurney, and Co. at the end of 1860 and the commencement of 1861, and possessed of a little penetration, could see that the "Corner House" was in great difficulties. They were trying to realize the securities they held, at no matter what sacrifice, especially when they knew that the losses would fall on their own shoulders. About this time I was constantly receiving the most inconsistent letters from them. I could never make out what they desired or what they intended to do. These letters betrayed much confusion of mind and great weakness of position on the part of the writers. Now they asked me to help them by turning the Greek and Oriental Steam Navigation Company into a limited liability company; then they wanted me to sign letters which would bring a new capitalist in; again they ordered forced sales of my grain, and wished me to resell the two steamers which were lying unfinished in the yards of the builders, in order that they might have the use of the money I had advanced on their construction. (See Appendix, Nos. 1, 2, 3, 4, 5, 6, and 49.) As these letters were worrying me out of my life, and taking my mind away from business, I went off to Lombard Street, to see if some arrangement could not be come to.

Mr. Edwards was then a permanent fixture there from noon till four o'clock. His clerk, Mr. March, drew up the balance-sheet of 1860, as follows:—

BALANCE-SHEET, 1860.

Dr.

	£	s.	d.
To Creditors unsecured	24,827	16	7
,, Bills payable	56,805	11	2
,, Creditors secured	352,264	1	6
,, Balance, being surplus	5,424	4	6
	£439,321	13	9
Jan. 11th, 1860.			
Capital	40,295	18	10
Profits	21,571	1	7
	£501,188	14	2

Cr.

		£	s.	d.		£	s.	d.
By Cash at the Bank of								
London		£772	5	9				
,, Sir J. Lubbock's		142	18	1				
,, In hand		37	5	0				
						£952	8	10
,, Bills receivable						941	10	0
,, Debtors						44,822	7	3
,, Property						24,699	4	3
,, ,, held by Creditors						367,906	3	5
						£439,321	13	9
,, Balance brought down						5,424	4	6
,, Sundry charges						9,080	15	8
,, Commissions						37,177	7	2
,, Interest						10,184	13	1
						£501,188	14	2

In this balance-sheet £47,362 0s. 3d. figure as the commission and interest I had paid them that year. I could not help being every day more and more struck with the remarkable difference between the appearance and manners of Edwards and David Ward Chapman, and Henry Edmund Gurney and Robert Birkbeck. Edwards appeared thoughtful and worried, but still was cool and decided in his movements. Chapman's face had lost all its colour, and his eyes looked as if much sleep did not visit them of nights; the dictatorial tone had gone from his voice, and he seemed timid and tremulous. On one pretext or another he was always engaged in one of the small rooms, either with the official assignee or some customer. Henry Edmund Gurney and Robert Birkbeck met all who came to deposit or withdraw money or discount bills with their natural voice and manner; there was more energy than usual in their movements, but nothing else unwonted. It seemed to me at this time that there were two sets of men in this great firm—one who knew that water was already in the hold of the ship, owing to the holes they themselves had made, and which they were now trying hard to pump out to save her from sinking; another who managed the navigation, but were unhappily ignorant of the goings on below. I became certain of this as the letters addressed to me were not signed by Henry Edmund Gurney or Robert Birkbeck; they were simply written by Mr. Bois, the chief clerk, under the direction of Edwards and Chapman. I also observed that Henry Edmund Gurney would refuse to accept any deposit, no matter how large, if the depositor sought to get a fraction more than the rate of interest which the house was paying. This showed no suspicion, on his part at least, of a want of money.

On the 12th of January, 1861, D. W. Chapman asked me to signed the letter No. 2 in the Appendix. I refused, and, after a short conversation, left him. On the 19th of the same month, I received a letter (No. 6 in the Appendix) from Mr. Bois, telling me that until I signed Mr. Chapman's letter of the 12th, the house would make me no more advances. I must here observe that, after the retirement of Mr. Lascaridi, the relations between the Greek and Oriental Steam Navigation Company and Messrs. Overend, Gurney, and Co. were much changed. We were no longer in the positions of mortgagor and mortgagee. I paid to their representative, Mr. Henry John Barker, all the freights, and he paid all the bills and expenses of the Greek and Oriental Steam Navigation Company.

When Mr. John Preston told me now that Messrs. Overend, Gurney, and Co. were about to sell my steamers to Mr. Zachariah Pearson, I hurried to Lombard Street (it was the 11th of May, 1860, about twelve o'clock), and fortunately obtained an interview with Henry Edmund Gurney. I asked whether what I had heard was true. Incapable of uttering a falsehood, he replied, with some little embarrassment:

"Friend, it is true, and it is not true."

He then requested me to see Mr. Edwards about the affair. I refused to do so, and told him plainly how astonished I was that a firm like his should break contracts in that fashion. I said that I was not responsible for losses entailed by the acts of gentlemen like Mr. George Lascaridi, Mr. Henry John Barker, or Mr. Edwards, whom Messrs. Overend, Gurney, and Co. had appointed to finance the Greek and Oriental Steam Navigation Company.

My words made some impression, but the independence of my tone offended the mighty capitalist. During a crisis every feeling of humanity is silenced in the breasts of these money men. They expect the martyrs of their system to accept, not only with patience, but with joy, the tortures imposed. Their pride swells quickly to vengeance in case of opposition, or even if an explanation of their apocryphal doings be asked.

"Your fleet is not worth the sum advanced upon it," said Mr. Gurney.

"It is worth more than the money I received from you," I replied. "But, of course, if you are pleased to debit me with £30,000 bonus, £12,000 of Mr. Lascaridi's accommodation bills, £12,000 of our money retained by Mr. Henry Barker, and £20,000 lost by your forcing the sale of the grain, and many other items, the Greek and Oriental Steam Navigation Company will very soon represent a loss absorbing all the profits which I have made with the steamers. However, supposing for a moment that the steamers are not worth the money advanced on them, the blame does not lie with me, Mr. Gurney, but with those appointed to act for your firm. I cannot be blamed by anybody." I finally protested against the proceedings of his firm, and declared that if they persevered in carrying out their intentions they should abide by the consequences.

"We can chop you, friend, in a few moments," he said, angrily.

"You cannot," I retorted. "Try, if you like; but remember that the public shall know the whole affair. The blame will fall on you, not on me."

Mr. Gurney, seeing that I was determined to die fighting, asked me to wait a few minutes. He passed into the adjoining room, where Chapman and Edwards were. The door was only partially closed, and the voices of the speakers were loud. I heard Chapman say I was a d—— f——, and, cursing the day he met me, exclaim—" Go you, Edwards, and speak to him yourself."

The ambassador entered the room where I was sitting. Noble simplicity, goodness, frank confidence —blended with prudence—were mirrored in his smiling face. The official assignee commenced by professing a sincere friendship for me. He then assured me that what Messrs. Overend, Gurney, and Co. projected doing was solely for my benefit. They had found a good buyer for half the fleet, and they considered that, under the circumstances, it would be well to relieve me of half my burden. It was in vain I endeavoured to show him that by disposing of my large steamers he would break up the line, for my small steamers in the Danube, which were the feeders of the large ones, would then be useless; with a less number of steamers than I possessed, the difficulty of paying my great debt would be increased. My arguments were useless, and no wonder. In fact, Messrs. Overend, Gurney, and Co. had that very morning prepared the transfer, as mortgagees, of my best steamers to Mr. Zachariah Pearson; Edwards showed me the bills of sale. I was thunderstruck, the more particularly because I saw in the list of the transferred vessels two of my best steamers. When Messrs. Overend, Gurney, and Co. did this, it appeared they forgot that the bill on which I paid a bonus of £30,000 was to be renewed. Mr. Edwards now stepped in to arrange matters. He proposed that two large steamers should be returned to me in exchange for two others. He also suggested that if I would sign the bills of sale for the six steamers, Messrs. Overend, Gurney, and Co. should give me a letter, engaging not to sell any more of my vessels as long as I paid them a fixed annual sum, until my debt to them was cleared. The miser may perform ten acts of charity; must we, therefore, affirm that he is charitable?

What time would it not take to recount all that took place between me and Messrs. Overend, Gurney, and Co., their brokers, agents, solicitors, favourites, and courtiers, before the day of sale arrived!

If I had hitherto been frequently deceived, it was not because I was unable to discriminate characters, but because of my inexperience in business, and because of a certain *laisser-aller* incidental to all men whose energies are overtasked. My eyes were now wide open. I saw the position of Messrs. Overend, Gurney, and Co.; I saw they were running a suicidal race. They were surrounded by men who plundered them, and yet such was their blindness, that what was very plain to the eyes of others seemed hidden from theirs. But let me tell you, reader, how my steamers were sold to Zachariah Pearson—a transaction which, within twelve months, entailed considerable loss upon Messrs. Overend, Gurney, and Co.

CHAPTER XXIII.

ZACHARIAH PEARSON.

Mr. Zachariah Pearson had been, a few years before, merely master of a vessel. By honesty, industry, and ability, he made money and became a shipowner, and afterwards a steam-owner. He was elected Mayor of Hull. A respectable career was open before him, from which, in an evil hour, he deviated, and attached himself to the train of Henry Barker.

Henry Barker, like another Dædalus, raised our poor Icarus so high, that the sun dissolved the wax that attached his wings, and he fell into the depths of the sea. Poor Zachariah Pearson had bought his steamers on credit, and when his acceptances fell due, Henry Barker provided the funds, by discounting his new acceptances at enormous interest, with bonus and douceur. Mr. Barker's mode of proceeding was this: He mortgaged the steamers in question to different capitalists charging one-half per cent. as commission, taking for himself an additional mortgage, because of the risk he incurred in drawing the bills, although it was agreed between him and the discounter that he should never be called on; besides that, the property was more than sufficient to cover the sum lent. The Overends had already advanced a round sum to Zachariah Pearson on the steamer Chersonese and her Government freight, and they now sold him my steamers, taking in payment his bills and a mortgage on the steamers. Mr. Barker had a commission of £250 or

£500 on the transaction, charged to my debit, for his trouble in drawing the bills.

I shall here briefly note down the chief causes that tended to produce this lamentable catastrophe. In the first place, the great commercial storm had considerably reduced the deposits in the great "Corner House." The money advanced by Messrs. Overend, Gurney, and Co. to shipowners, railway projectors, and other speculators, was the property of the depositors, and not their own money. Now, the money so advanced was not only locked up, it had the additional disadvantage of being locked up in property which, at the best of times, could not easily find a purchaser, and which, under the actual circumstances, was wholly unmarketable; and superadded to these difficulties was the perplexity arising from the fact that in the majority of cases the money advanced was much in excess of the value of the security. In what way could the bankers meet the demands of the alarmed depositors, who clamoured for their money? But one way remained, which was to discount with their indorsment the big bills of the owners of the steamers or other property that they held. At least, this was the mode in which their legal adviser, the great mathematician, Mr. Edward Watkin Edwards, proposed to finance them. It was the application of his system to the heads of the firm, as he had already tried it on those who had had the misfortune to fall into their power. As the entire financing department was under the control of this distinguished personage, his authority was absolute as that of the Olympian Jupiter. Even the Titans, Overend, Gurney, and Co., were under his thumb.

My account, owing to large credits given me by Messrs. Overend, Gurney, and Co. to buy grain, with bonus and commission thereon, exceeded, at that time, £300,000. The great Necker of the establishment advised, in order to finance both the firm and me, that the grain should be sacrificed coûte qu'il coûte, and turned into cash. He also recommended that some of my steamers should be transferred to an account where the acceptances could be more easily negotiated than those of the Greek and Oriental Steam Navigation Company, of which the representatives of the Lombard Street house already held so many.

Mr. Zachariah Pearson had, at that time, no steamers mortgaged to Messrs. Overend, Gurney, and Co. except the Chersonese, which was already paid off. Knowing that his credit was good at Hull, with Smith Brothers, his bankers, the great capitalists sent for and asked him to buy a portion of my fleet.

"I cannot," said Mr. Pearson.

"You will not be asked to pay any cash," said Edwards; "you need only give your acceptances."

"I am already overworked," said Pearson.

"It will be a good thing for you, Pearson," said the official assignee.

Mr. Pearson still hesitated; in fact, he did not like to enter into the transaction at all. They, however, talked him over, and he at last agreed to meet Edwards at the office of Messrs. Crowder and Maynard, the solicitors, in order to complete the transaction at once, as Messrs. Overend, Gurney, and Co. were most anxious to have the bills that day. On the way to Coleman Street, Mr. Pearson began to change his mind, so that when he had arrived at the solicitor's office, he told Edwards that he was not at all inclined to carry out the transaction. On hearing this Edwards looked terribly disappointed, and after a short silence said, "Pearson, come and speak with Mr. Gurney." On their arrival at Lombard Street, Pearson went into the little room facing Birchin Lane, whilst the adviser went direct to Mr. H. E. Gurney to give him counsel in the matter.

After the lapse of a short time both the capitalist and the disposer of the capital entered the room where Pearson was.

"Now, friend Pearson, what is the matter again?" said Mr. Gurney. A further debate on the subject ensued, and after a short time they brought Pearson round again, and then Mr. Gurney said to him: " Friend Pearson, I will give you, on the spot, a cheque for £500, if you will sign the acceptances and complete the business to-day. Bois, draw a cheque for £500 for Mr. Pearson."

The cheque was handed to Mr. Zachariah Pearson. The official assignee gave him his *word of honour* that he had nothing to fear on account of this noble transaction. Mr. Henry John Barker, the favourite, *secundus*, of Mr. David Ward Chapman, was also regaled with a cheque, in compensation for his trouble in drawing the bills of exchange. Edwards and Pearson returned at once to the lawyers', where he signed the mortgages and accepted the bills. Edwards then took the acceptances, and, like a Don Cossack, started with them for the " Corner House," where arrangements had been previously made to discount them with some melting-house.

"This is my death-warrant," sighed poor Pearson.

"Not at all. It's all right," said Edwards, his smooth face irradiated with a smile.

Oh, glorious English " All right!" How multitudinous are your meanings—how complex your significations!

Scarcely had Pearson left the lawyers' when Edwards hastened after him.

"Hallo! I say, Pearson," cried out the smooth-faced worthy, "what are you going to give me for this capital job I've done for you?"

"Capital job!" said Pearson; "I call it my ruin. I have no trade to employ all these steamers. It will be my ruin, I say."

"Not at all, not at all. All right, I tell you. You have to thank me for putting you in the hands of Overend, Gurney, and Co. They will make you one of the greatest men in England."

This modern Rhadamanthus called into play all his eloquence and logic to prove to Zachariah Pearson that his share of the transaction was worth £1000.

Poor Zachariah Pearson! I have it from his own lips that the same day he gave Edwards a cheque for £1000, double the bonus he had himself received. Verily, verily, Mr. Edwards knew how to trim his sails so as to catch the wind from whatever point it blew.

The intrigues of that "Corner House" were more than fiction would dare to fancy. As a sequel to the above transaction, I shall here mention that Mr. Edwards pressed, and even worried, Mr. Pearson to take a brother of his (Edwards's) into partnership. Entangled as Pearson was in the Lombard Street meshes, he resolutely declined this honour, exhibiting in his refusal more determination than I had shown on a somewhat similar occasion.

Whatever commissions the above-named two *gentiluomini*, Barker and Edwards, were paid for the job, were added, of course, to the price of the steamers.

It is said that Bishop Roquette was the original of

Molière's *Tartuffe*. Can any one tell me, in this Lombard Street comedy, amongst so many, which is the real *Tartuffe*?

"What does all this mean?" I said to my book-keeper, Ross, when, having returned to my office in despair, I told him of the breaking up of our line.

"It means," he said, looking up from his desk, and speaking in a low voice—"it means, perhaps, that they want money themselves."

I asked whether he had any grounds to think they did want money.

"Yes," he said; "it's the opinion of Mr. March, Edwards's confidential clerk. In making up the books this year, and going through the accounts, he came to the conclusion that the firm is in want of means—they are not right. But, pray, do not speak of it; do not mention my name."

I left the book-keeper's room silent and almost petrified. I found Mr. John Preston in my own room. I told him of the breach of faith committed by the Gurneys in selling my steamers. I also repeated what I had learned from Ross. Mr. John Preston was deeply grieved at the breaking up of my line.

It was through Mr. John Preston I had bought and chartered nearly all my steamers. He is a man of high integrity and delicate sense of honour, combined with cool judgment and rare business capabilities. I have known him to work out simultaneously the interests of three antagonistic parties, and, without offending any, satisfy all. The steadiness with which he resists, and the skill with which he parries any attempt to *pump* him with regard to your opponents' affairs, afford the best security that he will be equally taciturn with regard to yours.

When I had told Preston what I had learned from Ross, he said:

"What a pity! what a pity! They have ruined Pearson as well as you. Pearson has not trade enough for these steamers. The Gurneys will be obliged, in a few months, to take them back and return them to you."

"Do you suppose," I replied, "that I would take them? Do you think I would submit to such child's play? It will be known to-morrow, all through the City, that my steamers are sold. The Gurneys have destroyed the credit of the Greek and Oriental Steam Navigation Company. Who can trust them now? They have made me pay £30,000 on these steamers. It may suit them to-morrow, for the sake of a bonus, to transfer the remainder of my steamers to another person whose bills can be more easily discounted, and then, when they will have stripped me of everything—left me absolutely without property—then they can pounce on me for the deficiency, and fling me into the talons of Edwards at Basinghall Street. Oh! no, no."

"Stay, stay," said Preston, with his habitual coolness. "Try to induce them to enter into a contract not to sell any more of your steamers, and to give you credit to build others."

"Induce the Gurneys to make a contract! Do you know they don't give one time to say a word?"

"Speak to Chapman."

"No. He hates me, because he knows I am a man with a tongue in my head."

"Speak with Gurney, then."

"I firmly believe he does not know what Chapman and Edwards cook. He and Birkbeck are in the dark about everything. Besides, when I go to speak with him, he sends me to Edwards. Edwards! If Edwards were Briäreus, with his hundred hands, he would not be able to manage the Galway, the Millwall, the East Indian, together with the Greek and Oriental and so many others."

I was fearfully excited. Preston could, of course, afford to be cool.

"I'll tell you what to do," he said. "Go to Mr. Edmund Gurney before the mortgagees transfer your steamers to Pearson, and try to make an agreement. Look here. Write a kind of circular, something like this."

Mr. Preston took a pen and sketched the following circular, which I gave to Turner, one of my private clerks, to copy:

The Greek and Oriental Steam Navigation Company.
10, *London Street, Fenchurch Street,*
London, 14*th May,* 1861

Taking into consideration the exteme dulness of business, and particularly the great falling off in the shipments of goods to the Mediterranean, we have withdrawn the following steamers from the line, viz. :—

Smyrna, Patras, Petro-Beys, Lord Byron, Admiral Kanaris, Modern Greece.

But it will be maintained with punctuality and efficiency by the following ships :—

Palikari, Scotia, Asia, Mavrogordatos, Leonidas, Zaïmis, George Olympius, Colletis, Colocotronis, and Rigas Ferreos.

Thanking you for past favours, and soliciting a continuance of your patronage,
           We remain,
              Your obedient servants,
          (Signed)   STEFANOS XENOS.

"Tell Edmund Gurney you consent to the sale of

the steamers, provided Pearson pays a good price for them, and that, with permission of the firm, you intend to issue this circular. If they approve, it will be a kind of bond upon them in the eyes of the public. They will not sell so quickly again without considering the consequences."

"Considering the consequences!" I exclaimed, "What do insolvent men care for consequences?"

"Now, calm yourself," said Preston, "and go to Lombard Street. I shall soon be back with you."

I put on my hat, and took my way to the "Corner House." I was prepared for the worst. My anger was chiefly directed against Edwards and Chapman, who were at the bottom of everything in that office, and who always avoided speaking with their victims under pretence of pressure of business.

On my arrival at the wonder-working mansion, I was shown, as usual, into one of the little rooms. I asked for Mr. H. E. Gurney. He was engaged just then, but made his appearance in a few minutes. He was evidently in better humour than he had been at our last interview.

"Well, friend Xenos," he said, "how are you?"

"I suppose, Mr. Gurney, my steamers are now sold."

"Yes, friend."

"At what price?"

"At a better price than you paid for them."

"I bought them a bargain. They are worth now 25 per cent. more than I gave for them. A property that brings me in 30 or 40 per cent. per annum is, of course, worth much more than I gave for it."

"Why don't you say at once," put in Mr. H. E. Gurney, sharply, "that your steamers are worth £100,000 each?"

"What is to be done now, Mr. Gurney," I said, "with this business of mine? As it is, it can never be made to pay."

"On the contrary, now that you have a smaller number of steamers you will be able to manage them better."

"That is not the nature of the line. Besides, you have left me more small than large steamers. And what will the world say? This is the *coup de grâce* to the credit of the Greek and Oriental Steam Navigation Company."

"Not at all. A man may sell horses that he has used, and afterwards buy others."

"Well, will you give me a letter to the effect that as long as I pay you £20,000 or £25,000 per annum against my debt to you, and pay 5 per cent. interest, you will not sell any more of my steamers?"

"Yes, friend;. I think we can do this."

"I have prepared this circular, which I intend to send round to the shippers."

I handed him the circular. He read it, and, having made some trifling corrections, returned it, saying:

"You may send it; but stay a moment."

He went out to speak to his Solon, Edwards. After a few minutes the great legislator entered. His smooth, polished face, lustrous with smiles, seemed to shed a radiance over the dimly-lighted room.

"How are you?" he said to me. Mr. H. E. Gurney tells me you have consented to sign the bills of sale."

"Yes," I said.

"Oh, you will be all right soon. You showed a great deal of temper though. You must give up temper in this office—you injure yourself by it. These people like you. They took you by the hand. They have the best intentions towards you. Believe what I say. You will soon be in a position that dukes and earls will envy."

I remained silent, not, I confess, without an effort. I thought of the Greek proverb : " The bread is another's, the teeth are his own."

" Will you come with me and sign the bills of sale at Menard's, the solicitor's ? " he said.

I consented. I found the bills of sale already made out in the names of Overend, Gurney, and Co. ; that is to say, they, as mortgagees, were transferring my steamers. The solicitors had now to make out new bills of sale, which I signed the following day. Edwards and Chapman were well pleased that I had done my part without further noise.

When, the next day, I went to the Lombard Street office, Mr. David Ward Chapman condescended to salute me with an affable smile. He told me that, now that I had signed the bills, he would do something handsome for me. I laughed in my sleeve as he spoke in that strain. Poor Chapman, who measured men by the length of their purse, not by their intellectual capacity ! Had he known that I was aware of the real position of his mighty firm—that I knew upon the brink of what a gulf Overend, Gurney, and Co. were standing—he would not have been quite so ostentatious of his patronage.

Reasoning from what I had learned behind the scenes of the great Lombard Street stage, I came to the conclusion that the mystery of the plot was this : Firstly, that not alone was the house insolvent, but that, secondly, the secret was known only to David Ward Chapman and to Edwards. I believed the other partners were wholly ignorant of the true state of things. I was led to this belief by observing that Chapman, who was living in great luxury, was thrown into a state of terror whenever the firm was called on for large payments; and when the smallest cloud appeared in the commercial sky, Edwards, spite of his gout, jumped out of bed, and busily applied himself to business. The Gurneys and Birkbeck, on such occasions, though they worked with increased energy, never exhibited the fear or uneasiness that would imply a sense of imminent danger. Certain it is, they yielded implicit obedience to the wishes of their pilot, Edwards. Captain, owners, officers, and crew of the once seaworthy ship, were as nothing before him. He was the sole navigator; he alone knew whither she was drifting, and in what port she was destined to find refuge.

When I returned to my office on the day that Mr. H. E. Gurney consented to my sending a circular to the shippers, I found John Preston there. I showed him the alterations made by Mr. Gurney in the circular. Mr. Preston having read, breathed on the interlineations, which were in pencil, and, handing me back the paper, said:

"Keep this; it may one day be useful to you."

He then asked: "Have you got a letter from them, promising not to sell the remainder of your fleet?"

"Not yet," I answered, "but I expect to have it. Mr. Gurney promised it."

"Try to get it as soon as you can. I do not think they will sell any more vessels; they won't wind up the line. If they are insolvent, as March said, they will be compelled to keep it for their own sakes."

CHAPTER XXIV.

WHO IS ENTITLED TO A COMMISSION?

It was a combination of circumstances that afforded Mr. Henry Barker an opportunity of smuggling Mr. Zachariah Pearson into the " Corner House." By this *coup* he not only gained a handsome sum as commission, but he freed himself and his friends from many risks and responsibilities which he had incurred conjointly with Zachariah Pearson. All these he now threw upon the back of Messrs. Overend, Gurney, and Co.

Previous to this time Zachariah Pearson had had but few transactions with Messrs. Overend, Gurney, and Co. These mighty, but misled Overends! Regarding matters in a financial light, they thought they were acting prudently; but, in their ignorance of shipping business, they were committing a double murder. They were assassinating the Greek and Oriental Steam Navigation Company, which they thought they were saving, and they were driving Zachariah Pearson to his destruction, whilst they thought they were doing him a service. It was in vain I tried to enlighten them upon my own affairs; in vain I told Mr. H. E. Gurney freely my opinion. Often in those days did the invocation of Voltaire recur to my lips—

> Descends du haut des cieux, auguste vérité,
> Que l'oreille des grands s'accoutume à t'entendre.

My line consisted of two classes of steamers—small

vessels, that were feeders of the large; and large vessels, that were transporters of the goods. Messrs. Overend, Gurney, and Co. transferred six of the large steamers to Zachariah Pearson, and left me, in the Danube, a fleet consisting almost entirely of small vessels. In other words, they took the large wheels out of the machine, and left only the small ones to work it. Here, then, was Zachariah Pearson, who had not a regular line, but merely ran his steamers on simple jobs to any sea—here was he, I say, at the head of an armada far larger than he could find work for, pay, or manage. And the advice upon which these changes were made was given for the sake of securing a few thousand pounds' commission. And what was the result? It was not long before Zachariah Pearson saw his real position. He determined upon a *coup de main* that would at once seal his fate. The American War was then at its height. He resolved to run the blockade of the Southern ports. Some of my former vessels were to be employed in this service, and Mr. Henry Barker's brother was appointed admiral of the flotilla. It was a mad project. Some of the vessels were too small to cross the Atlantic, others were of too mediocre a steam power, and some others, when loaded, drew more than seventeen feet of water. The attempt was a signal failure. Some of the vessels were captured, others were stranded, and poor Zachariah Pearson, driven to bankruptcy, was stripped of his last penny by his pretended benefactors. The estate of the Greek and Oriental Steam Navigation Company lost £2000 in this bankruptcy, because Mr. Henry Barker, when he was appointed to finance it, exchanged accommodation bills with Zachariah Pearson for £3000, and gave us the following letter as security for some shares of the steamer Emmeline. But when Zachariah Pearson failed the steamer Emmeline could not be found—she had been either lost or sold long before—so our security was valueless.

Russia Chambers, 98, High Street,
Hull, April 23rd, 1860.

To the Greek and Oriental Steam Navigation
   Company, London.

 Dear Sirs,

  We enclose herein our acceptances for £3000, and, as security for their payment, we hold for your disposal, should they be dishonoured at maturity, which they will not, eleven sixty-fourth shares in SS. Emmeline.

We are, dear Sirs,
Yours truly,
For Z. C. Pearson, Coleman, and Co.,
Thomas E. Johnson.

But the comic part of the story remains to be told. At what may be called poor Pearson's commercial funeral, Mr. Henry Barker appeared amongst the chief mourners. *He was a creditor to the amount of* £15,000. However, he had one consolation; he was amongst the *secured creditors.* This was not bad on the part of a man who, three years before, had been obliged, for private reasons, to hold the furniture of his house in the name of his aunt. Ultimately Mr. Henry Barker was paid his £15,000 in full. Mr. C. E. Lewis, before the Commissioner of Bankruptcy, Goulburn, on the 11th of March, 1864, pronounced a long condemnatory address against me; he finished by saying:—

 It is the key to the whole of Pearson's subsequent misfortunes. Although, 'probably, he had steered through many shoals on the high seas, he did not know how to steer clearly through the shoals which sometimes exist under the sea of prosperity. He had this opportunity of buying this fleet of ships for £80,000, upon terms of credit. Dangerous facility! especially if you consider the terms on which he was allowed to buy them:—" Oh, you need not pay for them; we will give you eighteen months or two years to pay for them, provided you give us bills, which we will renew every three months." What a temptation that, to a man who had made £20,000, to launch out so as to be the largest ship-owner in Hull; in point of fact, to be master of a large fleet of ships! He succumbed to it. A few months afterwards the convenience of the arrangement—at all events, to Messrs. Overend and Gurney—was obvious; they got rid of Xenos and Co., the previous mortgagors—gentlemen so well known in the City of London as always in the courts of law. There was the Patras, the Indian Empire, the Modern Greece, and all those ships that we have heard of in connexion with Stefanos Xenos and the Greek and Oriental Steamship Company. Mr. Pearson arranged to pay for them by bills renewable every few months; and, a year afterwards, this seemed so convenient an arrangement that £66,000 worth of ships were sold to the bankrupt by Overend and Gurney upon the same terms.

The COMMISSIONER: Did Overend and Gurney sell him the ships?

Mr. C. E. LEWIS: Yes.

But he never divined the means by which Zachariah Pearson had been ruined. Nor were the public wiser than Mr. Lewis; though they talked a great deal about the affair, they did not know that the men who contrived to throw the blame on others managed to put the profits and commissions into their own pockets.

It was after my steamers had been transferred to Zachariah Pearson, and the account sent to the Greek and Oriental Steam Navigation Company, that I learned from Mr. Henry Barker that he had received a small commission for his trouble in drawing the bills; at the same time I complained to Edwards about these commissions, saying I knew they would be tacked on

to the steamers. Without mentioning to me that he
also had received a commission, he answered coolly:
" Every one is entitled to his commission. I take
commission, Barker takes commission, the Gurneys
take commission, you may take commission. We
cannot work for nothing."

CHAPTER XXV.

MY FOURTH FINANCER.

THOUGH the record of the remaining term of the Greek and Oriental Steam Navigation Company will be short, it may be found amusing, perhaps instructive. It is like the story of the decadence of an heroic little state, which, after accomplishing some illustrious deeds, falls into a condition of slow decline; then ensue terrible confusion and anarchy, redeemed by occasional acts of prowess, until the fatal day arrives when death's dark pall covers everything, and even the name of the country is forgotten.

When Messrs. Overend, Gurney, and Co. transferred six of my steamers to the name of Zachariah Pearson, they, no doubt, thought that I might imitate their breach of faith, and take advantage of what property I had in my hands—freight, grain, or steamers —to protect myself; consequently, they decided on sending a nominee at once to my office. Mr. David Ward Chapman was the first to announce this happy decision to me, by informing me that the coming man was the best I could have for an associate.

"What," said I, "another financer!"

"You will like him very much," said Mr. Chapman. "He is very different from Messrs. Henry Barker and George Lascaridi."

Mr. Edwards, shortly afterwards, preached me a long sermon regarding his good feeling towards me, and his good intentions towards the Greek and Oriental

Steam Navigation Company. Finally, he informed me that the new comer was Mr. G. B. Carr, who had formerly been a merchant in the Levant trade.

I said nothing; I expressed neither satisfaction nor dissatisfaction; it was no use to speak to them. I had formed a fixed opinion regarding these two gentlemen, so I immediately sought an interview with Mr. H. E. Gurney. He was rather cool to me on account of what had occurred when last we met. I complained to him of the introduction of this new friend of Messrs. Chapman and Edwards, and I also strove to explain certain matters to him, when he interrupted me abruptly by saying—" If you don't like it, go out altogether."

" That is not so easy, Mr. Gurney," said I; and added—" Is this what you call justice?"

" Friend Xenos, you do not know yourself what you want. Friend Xenos, do you know who is your greatest enemy in this world? It is Mr. Stefanos Xenos. There is not a duke's son who would not envy your position, and yet you always have something to complain of. Go, and be satisfied. The gentleman whom we are going to send you is very different from the others."

An interview with a great capitalist must necessarily be short, as his presence is constantly required by others, and particularly when he leaves the care of your affairs to a devoted partner and his legal adviser.

A few days after this interview, Mr. G. B. Carr made his appearance in my office, bringing with him a letter from Mr. Edwards. (See Appendix, No. 14.) As this letter is dated the 9th of May, 1861, the very time at which my steamers were transferred to Mr. Pearson, it is very evident that when the official assignee prepared this little *coup de main*, to prevent me from imitating their breach of faith, he simultaneously chose a representative. It was notified to me now that they would pay Mr. Carr £500 a year from their own treasury. Messrs. Chapman and Edwards, of course, judged men's characters by their own—ἐκ τῶν ἰδίων καὶ τὰ ἀλλότρια. I may as well add here that, although Mr. Chapman told me that his firm would pay the new financer, such was not the case; and so the Greek and Oriental Steam Navigation Company had to pay double—viz., £1000 per annum.

Mr. G. B. Carr, of No. 5, Laurence Pountney Place, is one of the most respectable men that I have ever met. He was once a Levant merchant, and engaged in large trade, but I am afraid that his goodness of heart was the means of bringing him to grief. Mr. Henry Barker had, I believe, a hand in that catastrophe too.

Mr. Carr was a man of large mind, whose eye gave abundant evidence of the sensibility of his heart. He was a close thinker, cautious in act and precise in words, calm-tempered, and unaffectedly religious. He was no longer young, but his activity was remarkable, considering his age. Had he been appointed two years before to superintend my finances, instead of Edwards, George Lascaridi, or Henry Barker, the Greek and Oriental Steam Navigation Company would now be one of the best of the Mediterranean lines, instead of being reduced to ruins, that served to others as stepping-stones to fortune. How different was Mr. Carr's conduct to that of his predecessors in my office! He was sincerely desirous of serving both parties honourably; and, although he had his own business (in association with Mr. Hoare, and trading under the name of Messrs. Carr, Hoare, and Co.) to attend to, he would devote one hour every day to our affairs.

In this fresh reconstruction, one of the first conditions imposed on me by Messrs. Overend, Gurney, and Co. was, that the *British Star* should not in future appear as part of the estates of my steamers. They also resolved to strike a balance-sheet of the past, and commence a new leaf with the installation of Mr. Carr. This good work was begun by debiting me with an enormous sum, upon which I was to pay 5 per cent. per annum. Messrs. Overend, Gurney, and Co. then sent me the letter in which they promised to leave me only ten steamers. (See Appendix, Nos. 15 and 46.) It was not as I desired it, but better than nothing.

With regard to the *British Star*, I was not sorry for this new arrangement. Though I was still losing money by the journal, owing to prosecutions by the Austrian, Russian, Turkish, and Greek Governments, yet it now began promising to pay, and, with the Exhibition of 1862, the circulation would be increased. The political war against King Otho was just then raging fiercely, and it was most important that I should have full control over the paper, in order to aid the national cause, of which the *British Star* was the chief organ.

Mr. Carr was scarcely installed in my office when the price of corn rose rapidly. We gave Theologos and Carnegie, in Galatz, large credits on Messrs. Overend, Gurney, and Co., and we bought large quantities of grain. On these purchases the profits were great. The sun of prosperity smiled once more on the Greek and Oriental Steam Navigation Company, but it was only a momentary gleam. Mr. Carr's venerable countenance was radiant with joy. He thought with me that we had sighted the promised land. We were quickly deceived. Theologos unexpectedly arrived in London; Carnegie followed soon after. They had heard vague rumours of my complications with Messrs. Overend, Gurney, and Co.; they had seen the sale of my large boats announced, and had come to the conclusion that the small ones in the Danube would quickly follow, knowing as they did that the small would be useless without the large. Acting under this impression, they came to London, one after the other to learn what was going on.

Mr. Alexander Carnegie is a veritable Paul Pry in his way: he wants to know everything. Now, from 1857 to 1861—that is to say, during a period of five years—I was most careful not to allow any one, most particularly Carnegie and Theologos, to know the secret of my dealings with Messrs. Overend, Gurney, and Co., nor on what terms I was doing business with them. I gave my agents at Galatz, from time to time, credits to the amount of £20,000, and even of £40,000, on the Overends and other bankers, to enable them to purchase grain; but beyond the fact of these credits, they knew nothing of my transactions with those houses. These agents of mine had not the *entrée* into the "Corner House," nor did they know Edwards or Chapman even personally. When, about the middle of 1861, Alexander Carnegie arrived in London, he made the acquaintance of Mr. Carr, in my office. He then, for the first time, saw a representative of the mighty capitalists; he then, for the first time, learned that the authority in the Greek and Oriental Steam Navigation Company was divided. I thought it prudent to give him a partial confidence. I told

M

him a portion of the story. What I told was sufficient to disorganize the strict discipline that the service of steamers requires. Whilst I was speaking I saw a change come over the countenance of my grateful clerk. When I had finished he told me, in a frank and careless manner, that now the *coup de grâce* was given to the credit of the Greek and Oriental Steam Navigation Company at Galatz, and that for the future he could not hope to draw a £50 bill on the house, so that I should be obliged to furnish him, as far as I could, with credits on bankers.

CHAPTER XXVI.

REMORTGAGING THE STEAMERS.

MESSRS. Overend, Gurney, and Co. did not wish that the *British Star* should appear on our London books; but as the Greek and Oriental Steam Navigation Company had, by a circular (see Appendix, No. 7) guaranteed to the Levant public the existence of that journal up to July, 1862, I was obliged to enter into an arrangement with Messrs. Theologos and Carnegie, by which they were to charge sixpence or a shilling additional per quarter on their invoices, to counterbalance the annual loss entailed by the decree concerning the *British Star* with regard to our London books. I do not know whether Messrs. Theologos and Carnegie regarded this as an intimation that they might put on as many additional shillings as they pleased, but this I know, from that time forward all hope of realizing profits by the purchase of grain disappeared. In my attempt to repair the loss incurred by the *British Star*, I furnished my clerks at Galatz with the means of sending me the most comical invoices. The loss of the *British Star* was, *per se*, as I have already said, a mere trifle; but it so happened, in those disastrous times, that the journal was a mainstay to the steam company. Messrs. Overend, Gurney, and Co., not comprehending how intimately the two enterprises were blended together, took a very narrow view of the *British Star* as an ancillary power, and never understood the error they committed in ignor-

M 2

ing the journal. They ought, under the circumstances, to have submitted to the judgment of those who understood the working of the machinery. But it was characteristic of these gentlemen that they had and had not confidence in the men with whom they did large transactions. Their policy may be described as putting one corporal to watch another. They sometimes discussed the minutest trifles, and questioned the pettiest details; at other times they treated carelessly matters of serious gravity. They were often rash in their decisions, ofttimes remorselessly striking down an upright man, as frequently obstinately refusing to punish a detected rogue who had known how to worm himself into their good graces. Benevolent in some cases, they were tyrannical and dictatorial in others, and exhibited the latter qualities especially when the man who fell under their frown was of an independent character, and had a will and tongue of his own. Never accessible to reason, the Overends quickly conceived an aversion to any man who attempted, by cool argument, to point out the inexpediency of their orders. As the Greek proverb has it, these gentlemen, after being extravagant in the expenditure of the oil and vinegar, were parsimonious in the use of the salt and pepper. They were always in extremes; they knew no medium. To their eyes there were only two classes of men in the world—spotless angels and murky demons. To their understanding there were but two classes of events—volcanic disruptions and airy trifles. To their moral perception there were only two classes of deeds—perfect honesty and deliberate fraud. They seemed to regard men either as dwarfs or giants. The varied blendings, the infinite shadings of moral stature that lie between the two extremes escaped their notice. Under the guidance of men so constituted, commercial concerns could never hold the steady course which ensures ultimate success. But whilst they were themselves the slaves of their nominal agents, the long-standing fame of their house dazzled outsiders, and those who sought their aid perished by the same means that had wrought the ruin of the great "Corner House."

But I have digressed. I shall return to my Galatz agents. The grain latterly shipped by Messrs. Theologos and Carnegie, though bought in the upper part of the Danube, cost as much as other merchants paid at Galatz, instead of being purchased at three or four shillings per quarter less. This was giving an adverse turn to affairs. If I were obliged to pay, near the sources of the Danube, the same price for grain for which other merchants could obtain it at the mouth of the river, where was the use of my small steamers? They were, manifestly, a dead loss. Upon this point I shall only add that a sharp correspondence began between my brother Aristides and Carnegie. The reply of Mr. Carnegie (Appendix, No. 16) shows the nature of the subject. Within two years from this date my two clerks were owners of five steamers, plying on the Danube in the very trade for which I had originally started mine. (See Appendix, No. 40.) What is more surprising is, that their *magazziniere*—a M. Hadji Andreas—after passing eighteen months in their employment, had realized a fortune of £10,000. This was not bad for a gentleman who, a few months before he was honoured with the commands of my clerks, was a bankrupt.

But the most deplorable part of the transaction

was, perhaps, the mode in which the grain was several times sold in London. This was so bad as to drive not only me, but even Mr. Carr, to despair. It was done in this way. Immediately on the arrival in London of the bills of lading of a cargo, I handed them to Messrs. Overend, Gurney, and Co., and they handed them to Messrs. Coventry and Sheppard, the cornfactors. The Messrs. Overend then drew on Messrs. Coventry and Sheppard three months' bills for a considerably larger amount than my agents at Galatz had drawn on them; the reason for this increase being that the cornfactors' acceptances included the freight of the steamers, and insurance, profits, &c., &c. They then charged me with the discount of their bills, as well as with a commission for accepting the drafts of my agents. All this was a great accommodation for Messrs. Overend, Gurney, and Co., but ruinous to me. My grain was now as completely in the hands of the cornfactors as my money had been a little while before in those of Mr. Henry Barker. My property, in both cases, was made a source of profit and moral credit to other people. You, reader, must see that when a cargo is not in your own hands, you cannot treat freely with the brokers, and deal with the highest bidder; you must accept the price the holder wishes. My case was still worse. Messrs. Coventry and Sheppard had three masters, to whose commands they were bound to listen. There were the Overends, who, when money was dear and corn cheap, ordered the grain to be sold, *coûte qu'il coûte*, in the warehouse. There was I, who, when I had better offers from other brokers, protested furiously against such monstrous sacrifice of property. Then there was Mr. Carr, who, feeling the injustice done to me by the sacrifice of the grain, was perpetually going backwards and forwards between Messrs. Overend, Gurney, and Co. and Messrs. Coventry and Sheppard. The letters respecting these transactions, Nos. 28, 39, and 41 in the Appendix, will better explain this state of affairs.

All the labour, mental and physical, that I had gone through began to tell on me. Man is not a machine, and no vertebrate animal could work and think as I had worked and thought without feeling the effects. In fact, I had drawn long and heavy bills on my nervous system, and now they all fell due together. I became fractious, quarrelsome, obstinate —in short, unmanageable. I no longer recognized myself. I was in a state of constant irritation; I quarrelled with my best friends upon the slightest provocation. My overtaxed nerves were strained to the utmost limit of endurance. And no wonder. In this mismanagement of the grain alone £40,000 was wasted in England, and £40,000 in the principalities of the Danube. Some of the grain was warehoused, against my wish, then was blighted. The warehouse expenses, double commissions, interest, and material differences in a nominal sale, brought another loss.

But this was not all. I began to see very clearly that I was placed in a position of utter helplessness, as a cruel fate had inextricably interwoven my affairs with those of Messrs. Overend, Gurney, and Co., who were hopelessly insolvent. The transactions which I have just named, show that they were very hard pressed for money. Another transaction which I am going to mention will make the terrible position of the "Corner House" apparent to the most limited capacity. They began to remortgage the remainder of my steamers to Masterman's Bank, to the Bank of London, and others, and made my agents at Galatz, Messrs. Theologos and Carnegie, or Messrs. J. Hamburger and Co., of the same town, or Messrs. Sidney Malettass and Co., Mr. Henry Barker's agents at Smyrna, or anybody else whom they could find, draw the bills on these banks, giving their guarantee to Masterman and the Bank of London for taking up these bills before maturity, and charging me with the commission and discount. The bills thus obtained were endorsed by the firm and " melted." In fact they were galloping to perdition as fast as Faust was hurried by Mephistopheles, and they dragged behind every man connected with them. The letters Nos. 20, 47, and 50 in the Appendix are three of those that they made me sign.

CHAPTER XXVII.

A SECOND BREACH OF FAITH.

I AT last saw that it was useless for me to work so hard, sacrificing my health and the interests of my family. An opportunity occurred, of which I profited, to speak seriously to Messrs. Overend, Gurney, and Co., touching the prospects of my family in case of my death. One of my steamers, the Leonidas, was lost in the Baltic. I held the policy of insurance, which was for £12,000 more than Messrs. Overend, Gurney, and Co. had advanced me on the vessel. I asked Mr. Carr to propose to them that I should be allowed to set apart £10,000 for my family. I might have done this without consulting them, as I held the policy, and it would have been perfectly just. After some negotiations, I handed over the policies in good faith, to allow them to take their just proportion of the insurance.* They kept the balance, and passed it to

---

* In equity and strict honesty I only owed about £2000 or £3000 to Messrs. Overend, Gurney, and Co. on account of this steamer, as I can prove. On the 30th of March, 1861, I chartered the Leonidas—which was only 602 tons gross register—with Messrs. Pile, Spence, and Co., of West Hartlepool, for 1100 guineas per month. This charter I assigned to Messrs. Overend, Gurney, and Co., and during five or six months they collected the charter money, whilst the Greek and Oriental Steam Navigation Company paid all the expenses and the insurances of the steamer. In March, 1861, I mortgaged this same steamer to Messrs. Overend, Gurney, and Co., to induce them to guarantee £8000, acceptances of the Greek and Oriental Steam Navigation Company to the builders of

my credit in the general account. This was another breach of faith. To remedy this in some measure, Mr. Henry Edmund Gurney proposed that I should insure my life for £10,000, and that Messrs. Overend, Gurney, and Co. should pay the premium. I always had a great aversion to insuring my life. I was now compelled to try it. My solicitor recommended me to apply to the Union Assurance Company —a new company, to which he was legal adviser. I had a severe cold. I believe I was the first applicant to this new company. The whole force of the medical staff was brought to bear on me. During two hours I underwent a searching examination. My whole anatomy, physical and moral, was laid bare. I told them that on three several occasions I had spit blood, but not during the last fifteen years. At the end of my two hours' vivisection I retired. In three days I received a letter from the secretary, to say that the directors declined to accept my life. It was plain

---

the steamer, who, after having received half the price agreed on whilst she was under construction, refused to deliver her to me when finished, owing to the discredit thrown on my company by the above-mentioned circumstances, and the failure of Messrs. Lascaridi and Co. Messrs. Overend, Gurney, and Co. gave me the means of meeting the bills for £8000, but they collected the freight of the Leonidas, amounting to at least £6000, besides £16,000 or £17,000 for insurances. They refused to return me the balance, under pretext that the Leonidas was given as collateral security, which was wholly untrue.

The charter of the Leonidas and the extra insurance must, according to my calculations, have left a profit of at least £7000. Placed as I was in the power of Messrs. Overend, Gurney, and Co., and under such extraordinary conditions—paying, too, such heavy interest and commission, and the steamers yielding above 30 per cent. per annum profit—I made it a point always to insure each steamer for a few thousands more than she cost.

that the £10,000 risk and the vomited blood frightened them. I applied to another office. The doctor there seemed satisfied, but hearing that I had spit blood, he asked whether I had been refused by any other office. I said "Yes." This finished the conference, and within three days my life was declined by the second company. I leave you, reader, to imagine the impression that these refusals were calculated to make on the mind of any man. I firmly believed that I was booked for the other world. I made a resolve never again to submit myself to the examination of an insurance doctor.

I had still a few thousand pounds in my private purse. This sum was the residue of the profits on some cargoes of wheat I had sold in the commencement of the Italian War, to which I have already referred. I resolved to do some business on my own private account, and buy an estate, Petersham Lodge, near Richmond, which happened to be in the market. I did so, and Messrs. Overend, Gurney, and Co., for the first time, gave me a proof of their personal affection by lending me £1700 for three months, on security, to complete the purchase. Poor Mr. Carr, who advised them to afford me this little accommodation, travelled miles backwards and forwards between the two parties in negotiating the affair. The meanness, unfairness, and folly exhibited by the great firm on this occasion showed me that I could not go on much longer in conjunction with them. (See Appendix, No. 9.)

Rich, but narrow-minded men are often, by the course of events, placed in a position that gives them the direction of colossal transactions. As long as things go smoothly, these people seem to become their position, and really look the part of greatness; but when losses commensurate with the greatness of their undertaking occur, then the meanness of their nature discloses itself in acts which, contrasted with their position, make them appear ridiculous. This was the case with Messrs. Overend, Gurney, and Co. They attempted to avenge their losses on the very men who were trying to save them from being robbed. Mr. David Ward Chapman was squandering thousands of the depositors' money in luxury; his favourites were fast accumulating colossal fortunes; his sycophants were daily initiating some doubtful business; and, meanwhile, the timbers of the great Lombard Street vessel were coming asunder—she was fast drifting towards inevitable shoals and sand-banks. Mr. Henry Barker had precipitated Overend, Gurney, and Co. on to the terrible rock of accommodation bills, whence so many others had sunk into the Bankruptcy Court. Messrs. Overend, Gurney, and Co. had already remortgaged my steamers to the Bank of London and to Masterman's. Bills were drawn on these two banks, which Mr. Henry Barker discounted or negotiated with a trifling commission for himself. I had long been suspicious of this mode of doing business, which I foresaw could end but in one way. In fact, Mr. Henry Barker had so tied these three firms with accommodation strings, that the stoppage of one must necessitate that of the others. This catastrophe would have occurred two years earlier than it did, but that the mania for public companies averted the terrible conclusion. With regard to myself, I was not very manageable. I grumbled every time that I was asked to put my name to a bill, or sign one of their binding letters. An imperative necessity compelled me to yield, but I was dissatisfied with Messrs. Overend, Gurney, and Co., and they were dissatisfied with me. We were mutually desirous of separation.*

---

* Messrs. Lascaridi and Co., crippled so long, at last stopped payment in 1861, and Messrs. Overend, Gurney, and Co. must have lost considerable sums of money. Their estate passed into the hands of Messrs. Coleman, Turquand, Youngs, and Co., and these gentlemen at once tried to make the Greek and Oriental Steam Navigation Company debtors to the amount of £22,350 0s. 6d. (See letters, Nos. 25, 26, and 27, in the Appendix.) The reader will now see the reason why, in 1860, I refused to Mr. Edwards to sign the contract of the 1st of August, 1860, except the release of Mr. George Lascaridi from such claim should be inserted in our partnership deed. (See page 97.)

CHAPTER XXVIII.

A THIRD BREACH OF FAITH.

I HAD long been dissatisfied with the relations existing between the great money-dealers and myself, when an occurrence took place that deprived me of all confidence in them. You remember, reader, the Trent affair. Out of that business there sprang up a sudden demand for steamers as transports. For a moment it was believed that England was on the verge of a war with the Great Republic. I tendered two of my steamers, of 1200 tons each, for a time-charter, at 30s. per ton per month, and for four months certain, to take Government stores and ammunition to Bermuda. This would leave a net profit of nearly £10,000 for the four months. The bargain was almost concluded, when Mr. Edwards, to oblige Mr. Lyster O'Beirne, signs another tender for the brokers, Messrs. Thompson and Tweeddale, of 27 Birchin Lane (see Appendix, No. 17), and offers the steamers for a lump sum of about £4000 each. This, so far from being profitable to the Greek and Oriental Steam Navigation Company, would entail a heavy loss. The price of coal at Bermuda rose at that time £4 per ton, so that the price of fuel alone, out and home, would be £4000 on each. A time-charter with the Government is always safe, because they pay for the fuel and all other expenses, except wages and insurance. Edwards's offer suited the charterer better than mine. It was on the point of being accepted, and mine erased. I was driven to madness when I learned this news, and Mr. Carr was so exasperated that he asked permission of Messrs. Overend, Gurney, and Co. to resign his post. Messrs. Overend, Gurney, and Co. summoned Edwards, scolded the favourite, and justified me. Edwards thereupon sent me a letter, dated January 2, 1862, addressed to the Comptroller of the Admiralty. (See Appendix, Nos. 18, 19.) In this letter Edwards says that I was the sole owner of the vessels; but that, in consequence of illness, not being able to communicate with me, he had authorized Messrs. Thompson and Tweeddale, as mortgagee, to tender the steamers. It was too late and too bad. The quarrel between the mortgagee and owner put an end to both charter-parties, and I lost another £10,000 through two of Mr. David Ward Chapman's favourites.

This act of tendering the Palikari to the Government by Edwards, for the sake of putting a paltry commission into the pockets of his friends, was most arbitrary. A mortgagee has no right to charter a vessel without the consent of the mortgagor. There was only this difference in my case: Edwards was only the nominee of Messrs. Overend, Gurney, and Co., and these gentlemen had Mr. G. B. Carr in my office acting as their representative. I consequently held the "Corner House" responsible for the consequences, and wrote the following letter to them at once:—

19, London Street, Fenchurch Street,
London, 5th January, 1862.

MESSRS. OVEREND, GURNEY, AND CO.

GENTLEMEN,

The act of Mr. Edwards interfering in the management of my steamers, thereby seriously damaging even a second time my credit, is another breach of faith upon your part, and the *coup de grâce* that you give to this unfortunate burlesque, the Greek and

Oriental Steam Navigation Company. So I am determined not to go any further except a proper legal agreement be drawn up between my solicitors and yours, for the future operations of the steamers.

Gentlemen, I would refresh your memory to certain facts with which you must be well acquainted, but it appears you have forgotten.

When I first had the pleasure of knowing you, I had a position in this city, a large business, an extensive patronage of the Mediterranean merchants, and a very large credit. This business, with all the controversies, made and not lost money; but since that time the accumulation of sums, through debts and losses, caused by other persons appointed by you, my account with you stands such that, if I accept it, will take me many years to pay it off, working like a horse round the mill, and after all I should be without either agreement or security, and at the mercy of the circumstances.

The following facts are much more painful to me. I organized this line, and made a contract with the merchants to transport their goods. I then, for that purpose, bought and built 22 steamers —viz., 11 large and 11 small, the small being only for the use of the large. Mr. Carr sees clearly now that we have no steamers to carry out our engagements, and that I have lost above 10,000 tons of goods, equal to £17,000, these last three months, not having steamers, so that the shippers desert me every day. Why? Because you broke down first the credit of the line, and then the line itself, by having sold four of the large steamers, even without giving me any notice, which makes five of the small steamers useless, more especially as two large steamers of the number having been lost lately, and you refuse to replace them. When I paid you £30,000 a bonus in the January of 1860, it was verbally but honourably agreed, in the presence of Mr. Barker, Mr. Edwards, Mr. Lascaridi, and myself, that you must extend your loan of money to twelve months longer—viz., all the 1861; and so in the end of 1860 was arranged with Messrs. Lascaridi and Co. for giving you the letter that made that firm responsible for us for all the 1861. However, in the beginning of this 1861, without our knowledge, you advertise in the *Shipping Gazette* the total fleet for sale, through Messrs. Tamperly, Carter, and Co., my enemies, and by doing so you trumpeted in all the kingdom that our line is broken down, that you are the disposers of the steamers, and you receive in your office tenders, so that all ship-brokers believe they will buy the property for a bargain. At the same time you receive me with favour, and, although you supply me with funds for our daily wants, you never tell me one word for what you undermine me. Of course, only I myself know the losses and the discredit that I gained out of such proceedings. I do not say anything about my agony of mind and severe struggle, because that is nothing to you. At last, one fine morning, you sell six steamers. You judged that you made a profitable business. My books show that they are sold for less money than they cost, therefore making a heavy loss. But I come to another most serious question. Was it not you who forced me, through Mr. Edwards, to enter into a partnership bond with G. P. Lascaridi? Mr. Edwards came one night to my office, and, in the presence of Mr. March, said to me that the wish of Mr. Gurney is to do so. Are you not the parties who charged Mr. Lascaridi to finance the office? Why, then, was I to pay his accommodation bills of £12,000? Was it not you who appointed Barker to pay and receive our money? Why am I to be responsible for his balance due to us? True, I was obliged to acknowledge and sign such a pawnbroker's account by *force majeure*, because Mr. Barker had my payments in hand, and threatened me, if I did not do so, that he would not pay my bills that day. You recollect that I came personally to complain to you about that. Was it not you, gentlemen, who ordered my wheat to be sold, cost what it might, against my protest and express wishes? Why, then, am I to suffer the terrible loss? Was it not you who appointed Mr. Edwards to superintend our office, and report position and wants? He never came; he never knew what we wanted, although we paid him £500 a year. If he had acted as Mr. Carr now acts, this business never would be in its present position, and would have saved us at least £70,000.

Did you not detain the steamers in the Danube and in the Victoria Docks for six months, by refusing to give us the credit whereby we could load them? Meanwhile, how have I acted in all this confusion on the other part? I worked day and night to load the steamers—to make them to pay and not lose money—struggling, in this discreditable position, to keep the shippers with me—fighting honourably with my competitors, and also with many unknown and secret enemies—to prevent the robbery of the supplies of the stores, the carelessness and cheating of captains and agents; to economize and save every farthing; to protect your interests by keeping in constant repair the steamers, and not leaving them to go to pieces; to protect the interests of the builders; to deprive myself even of the necessary wants of my house, when I paid daily thousands to brokers; to mend large holes that others daily opened; at the last to allow myself to be in the present condition. Now, Mr. Edwards, your nominee or mortgagee, tries with others to make the best out of that by taking even the management out of my own hands.

Gentlemen, you understand that if my conduct to this moment has been honourable—and you are in a position to know it better than any others—I do not mean to proceed any further in this zigzag road. I am an open, sincere-dealing man, and I like straightforward ways. I cannot any longer repose myself under the sword of Damocles. We must have a clear procedure by contract, with certain and indisputable bases to work upon.

The bases that I ask are reasonable and just—

(a.) As long as I give you 5 per cent. interest for your money, and 5 per cent. against such money as you advanced me, not to sell any of the property.

(b.) If the business does not make in the end of one year any profit, you may at once sell or dispose of it as you like.

(c.) If, after paying you 10 per cent., there is a large profit left above £5000, you take half of it, and the remainder to be left to me to carry on the repairs and the business of the line.

(d.) The mortgage of the steamers to pass in the name of one of your firm, and not of any nominee.

(e.) The accounts to be regulated every six months, to prove if the business makes or loses money. The management of the steamers and their cargoes to be left entirely to me; and should other persons interfere by your authority, you to stand any losses that may occur by reason of such interference.

(f.) Whatever wheat is sold by Messrs. Coventry, Sheppard, and Co., and the contracts not acknowledged by me, the same to be returned by Messrs. Coventry, Sheppard, and Co. to me, or you to bear the loss.

(g). In case of my death, a gentleman to be appointed, according to my will, to go on for my benefit, as long as the business pays as above.

(*h.*) A balance of £10,000 to be left always at the banker's, to keep up the credit of the company.

(*i.*) A proper credit to be given to the house at Galatz for the working of the business. You may send a person there to superintend your interests.

j.) You can have in the London offices Mr. Carr, or any other gentleman approved of by me, to superintend the finances, advise, and see that every transaction is correct.

(*k.*) The four big steamers to be insured for £27,000 each, and the Powerful for £22,000. If any of the steamers are lost, and you will not let the money paid by the underwriters be employed in building another ship, then you take £25,000 and I take £2500.

(*l.*) You will give me £10,000, money due to me, if not for services, for the Leonidas, for which you were paid without being entitled to same, because your advances to me were only £8000.

(*m.*) All the wheat now in London to be kept in warehouse till the end of September. I to have the option of selling it by any broker who gives the highest prices, but Messrs. Coventry, Sheppard, and Co. to have the preference when they bid the same as the others; after that, you can sell it as you like.

Yours obediently,
STEFANOS XENOS.

I found the official assignee had two safety-valves by which he escaped when pushed into a corner by his own folly. In the first place, there was his wretched memory, that could not recall anything; and, secondly, there was his marvellous intermitting illness, that was sure to attack and confine him to bed when the *dénouement* of any misadventure, induced by some error of his, was about to take place. Under such circumstances, he was represented by some confidential clerk, upon whom the blame of the whole, transaction was thrown.

I had now begun to think seriously of freeing myself from the trammels in which I was bound. I was wasting my energies trying to fill sieves. This majestic house of Overend, Gurney, and Co., that dazed the eyes of the commercial community by its colossal transactions, was, in the estimation of those initiated in its *apocrypha*, an object of pity and contempt. There is a vulgar Greek saying that not inaptly represents the state of things in the great Lombard Street house in those days: Ἐμβῆτε σκύλοι στὸ μαγειριὸ χορτάσετε καὶ μὴ πληρώσετε ταῖς πορτζιόνες (come, dogs, into the cook's shop, and help yourselves without paying your portions).

Such was the confusion that prevailed at the "Corner House," that disorganization would be too mild a word to describe it. That great edifice was, so to speak, a mighty mercantile phantasmagoria that deceived the vision of the whole metropolis. Happy and favoured was the man, in the estimation of bankers and merchants, who enjoyed the privilege of *entrée* into that great commercial stronghold! But often those who were envied as favoured, were in reality victims—victims, too, who dared not complain. Nay more, these men even now, when the bubble is burst have not the courage to come forward and tell the truth. They are afraid to speak. Men who during eight years were tortured beneath the Lombard Street lash, have not now the social courage to come forward and tell the story—and why? "No banker would do business with us again were we to tell these transactions. Every merchant would shut his office door against us were we to speak of such things."

The standard of commercial morality must indeed be in a state of great depreciation if we have to be guided by such thoughts. Must we imagine that Diogenes has consumed a gallon of the best colza oil in searching amongst our capitalists and their agents for an honest man, and ultimately extinguished his lamp in despair?

CHAPTER XXIX.

CRIMINAL OR NOT?

My small Danube steamers being no longer backed by the large ones, had not sufficient employment; I therefore chartered several sailing vessels, and loaded them with grain at Sulina. This was a profitable speculation, but, unfortunately for me, the Galatz people, in this general anarchy, skimmed the cream for themselves. But the worst of the matter was, that though my agents drew on me at sight for the value of their cargoes, Messrs. Overend, Gurney, and Co. thought proper to be offended, and said I had been guilty of a breach of faith. Now, the truth was, I attached so little importance to the affair, that when the cargoes of the company were being sacrificed by Messrs. Overend, Gurney, and Co., I told Mr. Carr myself that I had got a better price for my private cargo by some shillings per quarter. It had come to this: Messrs. Overend, Gurney, and Co. found I did not suit them for their accommodation bills, and I did not choose to work perpetually for them and their favourites. I will give you, reader, an instance of the petty tyranny and jealous assumption of these people. A public company was at that time coming out—by-the-by, that same company afterwards became a great one—I was one of the directors; Messrs. Overend, Gurney, and Co. signified to Mr. Carr that I must withdraw my name from the direction. I was compelled to do so. (See Appendix, No. 37.) I am naturally independent-minded, and here I found myself the cat's paw of men who had no mercy on me, and who threw on me the blame that legitimately attached to the conduct of others. Their weight at that time with the English public was such that they could crush a giant. To break openly with them would be madness; to reason with them was impossible, for they would not grant me a sufficiently long audience. However, Mr. Carr saw how things were; he knew the truth, and he spoke out. Messrs. Overend, Gurney, and Co. were neither deaf nor blind. They had lost thousands through the intervention of their favourites, but a portion of these losses being put to my account looked to them like a gain. I saw that the crisis which was to separate them and me was fast approaching. I was determined that they should not crush me as they had crushed Zachariah Pearson and others.

On every occasion when breaches of faith on the part of Messrs. Overend, Gurney, and Co. drove me almost to despair, I used to rush off to the office of my solicitors, Messrs. Thomas and Hollams, of Mincing Lane Chambers, for advice. Mr. Hollams is the partner whom I generally saw. A more prudent and straightforward man there is not in the profession. He is possessed, besides, of immense talent, sagacity, and energy. In these interviews he would always strive to allay my excitement, and recommend me to do nothing rash. He would point out the inequality which existed between the two parties. I was weak, a foreigner, and in their power; whilst they were the giants of the City, and possessed of boundless influence. Could I have persuaded Mr. Hollams at that time that the "Corner House" was rickety to the very foundation, the history of the connexion between it and the Greek and Oriental Steam Navigation Company would long since have been in the hands of the public; but whenever I broached the subject to him he would smile, and give me no answer, probably thinking that the idea was so absurd as not to merit one.

At the time when Edwards caused the above-named loss, and consequent exposure and discredit amongst my friends, I happened to have in my possession, in cash principally, and bills of lading for cargoes of grain to arrive (see Appendix, Letters 21, 22), the sum of £100,000. Although my agents at Galatz, when buying grain, drew on Messrs. Overend, Gurney, and Co., the bills of lading were made out in my name, and sent to me direct. Driven to frenzy now, and consequently determined to break off all relations with the "Corner House," and to bring the whole matter before the public, I went to Mr. Hollams. I cannot at this length of time recollect the exact words I used in my interview with him, but I know that I told him, and not then for the first time, that Messrs. Overend, Gurney and Co. were insolvent, that they had taken possession of my property, and charged my account with every possible kind of figures. I then asked him, if I were to keep the money and bills of lading I then had in hand until I could get a settlement, and a satisfactory contract drawn up by our respective solicitors, could I be prosecuted criminally, or could they only bring an action against me at common law.

Mr. Hollams replied simply that he thought it very dangerous for me to carry so much money about me. He then placed himself opposite the fire, and, putting his head on his hand, fell into deep thought. After the lapse of several minutes of silence, I again asked him whether the act I contemplated was a criminal or a civil one.

"I do not think it is a criminal case," he answered; "it is an account current; you have an account current with them. But you will not proceed to such extremities?"

"I have no wish to do so," I answered; "but what am I to do with such men? The property is nearly ruined, and is becoming daily worse. I am nothing but an automaton in their hands for accommodation bills."

Mr. Hollams did not answer immediately, but he said at last, "Go to them, and talk the matter over."

"It is no use," I replied; "they have no time to listen to me, and will only send me to the inevitable Edwards."

"I cannot advise you to take the step you are contemplating," said Mr. Hollams; "go and try to arrange matters with them."

"Good!" I answered; but if they stop payment in the meantime, where shall I be after so much hard work?"

"You may chance that," said he, smiling, for he would never admit that I was right in thinking that the "Corner House" was insolvent.

"Well," said I, as I left, "I will follow your advice, for I will not give them an opportunity of saying a word against my character."

On my return to my office, I sent to Messrs. Overend, Gurney, and Co. the bills of lading, and the money received from the underwriters for the Marco Bozzaris. This was the fourth steamer which had been lost, leaving a profit of from £4000 to £5000. I may fairly say here that the Greek and Oriental

Steam Navigation Company made a profit in two years, on the insurances of the four steamers lost, and on the charter of the Powerful to the Government, of £50,000,* all of which was absorbed in this βορβορος of the "Corner House."

I telegraphed to Carnegie at Galatz, bidding him to come to London. I wanted his help for some months in the office. Mr. Carr being himself in business, was not able to afford more than one hour per day to my affairs. Carnegie came to London, but he did not seem pleased at being recalled and separated from his dear partner. No wonder—Galatz was a veritable Pactolus for the Scotchman.

Before entering into particulars regarding my breaking off with Messrs. Overend, Gurney, and Co., which occurred in the beginning of 1863, I shall recur to events connected with the *British Star*, which took place about the middle of 1862, and which contributed not a little to complicate my position.

* Admiral Miaoulis	£15,000
Powerful	15,000
General Williams	8,000
Leonidas	7,000
Marco Bozzaris	5,000
Total	£50,000

CHAPTER XXX.

POLITICS.

ABOUT the commencement of 1862, the tone of the *British Star* was so intensely liberal as to appear to many to be republican. Its denunciations against tyranny were as fulminating as those of Thomas Carlyle, in his "Republican;" but, if it advocated liberty, it was liberty based upon constitutional principles and Christian civilization.

The *British Star* was not what the Turkish Government accused it of being, when the Sublime Porte suppressed its publication. I have already explained the motives that induced me to establish that journal on the free soil of England.

It was about the close of 1861 that the *British Star*, by proposing Prince Alfred as successor to the throne of Greece, excited a general feeling throughout Panhellenium against King Otho. The very persons in different towns of the Levant who, at the first starting of the *British Star*, opposed the publication, now held public meetings to approve the policy of the journal. I received letters of encouragement and congratulation from every part of Greece. The press in the kingdom of Greece was shackled, and more than one editor shut up in the prisons of Carbola and Medressè. The Greek press in Turkey was chained under the despotic local government, and compelled to silence upon the wrongs and sufferings of Greece. King

Otho had recourse to what he believed an infallible policy. He tried to buy me. For this purpose he sent a Mr. Barozzi, one of his favourites, to London. This gentleman was empowered to offer me honours, appointments, if the *British Star* would only change its tone. I replied laconically, that the desired change might be obtained without any expenditure on the part of His Majesty, either of honours or appointments. King Otho had only to change his ministers, re-establish the national guard, appoint the successor to the throne, and not interfere with the constitution. Mr. Barozzi left London to join His Majesty at Munich. Within a few months the revolution at Nauplia broke out—an event which ought to have opened the eyes of the poor monarch to his real position.

The outbreak at Nauplia was regarded by King Otho's agents as a proof of the necessity of arresting the circulation of the *British Star* in Greece. They went so far as to open the letters in the public post-office, in order to discover with whom I was in correspondence. My unfortunate correspondents were immediately marked as conspirators. Beset in this way, I hit upon another plan. I printed upon thin paper the political articles of the *British Star*, and sent them enclosed in envelopes to Paris and Malta, to be posted in these places as ordinary letters. At the same time I wrote to the subscribers, promising that the remaining portion—the literary part—should be delivered to them as a bound book at the close of the year. These arrangements naturally entailed great expense. The copies of the *British Star* intended for the subscribers in Greece, I sent to Constantinople; there they were packed into cases having secret drawers, and so smuggled into Athens. Sometimes the cases were filled with bottles, in which the leading articles of the journals were safely corked down.

You are aware, reader, that during the Crimean War the inhabitants of Constantinople and the Levant were the last to learn the details of the campaign, and the information that did reach them was through the English press. The *British Star* was the only journal in which the blockaded soldiers of Nauplia had a chance of publishing an account of their *petite guerre*, and the events connected therewith. This increased the demand for the paper, whilst the patriotic spirit that breathed through its columns kindled the enthusiasm of the Greeks of Turkey, and excited their ire against King Otho. The Bavarian had recourse to a *ruse* that gave a fatal blow to the *British Star*. His mode of operation was this: He instructed Mr. Barozzi—the same who a few months before had had an interview with me in London, and who was now dragoman to the Greek embassy at Constantinople—to represent to the Turkish Government that the chief aim of the *British Star* was to induce the Greeks of Turkey to revolt against the Sultan's authority. Now, I had always avoided writing against the Turks, and except on two or three occasions, when their financial policy was alluded to, no mention of the Turkish Government appeared in the columns of the *British Star*. However, it so happened that at the termination of the Nauplian revolution, a number of Greek ladies and gentlemen— members of the most influential families of Constantinople—repaired to the island of Chalcis, and there crowned a copy of the *British Star* with laurels and roses. This patriotic demonstration Mr. Barozzi thought proper to construe as the germ of the Greek revolution. He hastened with the intelligence to the

Grand Vizier, and persuaded him that the cause of the bitterness exhibited by the *British Star* against the Greek Government was that King Otho would not sympathize with its revolutionary principles and projects. The Vizier, alarmed, fancied that the Greeks were on the eve of a revolution. He hurried to Sir Henry Bulwer. Sir Henry, without loss of time, telegraphed to Earl Russell. The noble earl—fond of adopting second-hand ideas, even in more important questions—immediately, and without examining into the truth of the accusation, gave orders to stop the transmission of the *British Star* through the British post-office at Constantinople. You remember, reader, the debate in the House of Commons on this subject in June, 1862. The circulation of the *British Star* being checked at Constantinople, was stopped altogether in London. But the arbitrary spirit of the British Government still further displayed itself when, a few months later, I applied to the Postmaster-General for permission to transmit through the English post-office another newspaper—the *British Beacon*—which I proposed to publish. I was informed that his lordship was not prepared to reply to my inquiry whether I might be permitted to forward this newspaper in the closed mails for Constantinople.

This fatal blow to the *British Star* cost me at least £3000 a year. The journal was stopped at the very moment when, after having entailed a loss of £15,000, it was beginning to pay by an extended circulation.

If the mass of the English people could read modern Greek, and study the columns of the *British Star*, they would be able to form some idea of the colossal fanatical prejudices—both religious and national—incarnated in the Russian party in Greece with which I had to contend. If the English people understood the struggles in which I became involved, and the sacrifices I made to propagate amongst the peoples of the East a knowledge of British institutions—to which knowledge was due the unanimous election of Prince Alfred to the throne of Greece—I feel convinced that, instead of being meanly persecuted, the *British Star* would have been loudly praised.

This affair of the *British Star* took place some few months before King Otho and his gracious queen disappeared from Athens. The reader, doubtless, remembers how quietly his expulsion was effected, or rather how easily his return was barred. King Otho was under a delusion. He fancied, because Athens continued tranquil after the revolution at Nauplia, that he was as firmly established on his throne as the most popular monarch in Europe.

King Otho was not a bad man, but he was wofully misled. He was not naturally a tyrant, yet he sanctioned or permitted the most tyrannical acts in his ministers, and so obtained the reputation of a despot. He was energetic and hard-working, but his energy was misdirected, and the time was wasted on trivialities which might have been employed in important affairs. King Otho would pass a morning laboriously spelling over the documents laid before him, whilst state affairs, involving the good of the nation, lay neglected. He was a moral man, but his was a squinting morality. He would never pardon a military officer or Government *employé* convicted of a love intrigue; but he winked at the known fact of his *aides-de-camp* being the patrons of murderers and brigands. He would frankly reinstate a wrongly-dismissed *employé*; but he would persecute unto the death the men—ay, and their families—who, supported by public opinion, openly and honestly opposed his political views and aspirations. He was ever dreaming of a Byzantine throne, upon which he was to mount by the aid of Austria and Russia. Otho was a proud man, and yet he was a mere toy in the hands of his flatterers, whose evil deeds were for the most part attributed to him. As I have already observed, he was misled. And yet, after the revolution of Nauplia, King Otho might have retained his throne, had he then thrown himself into the arms of his people, dismissed his ministers, and re-established the national guard. But, instead of this, he remained obstinately deaf to the cries of the public voice, so loudly raised against his policy, and he unwisely left his capital at the very moment that political changes were inevitable.

CHAPTER XXXI.

THE ELECTION OF PRINCE ALFRED.

I HAD intelligence in London of King Otho's dethronement three days before it was known to the English public. Although the *British Star* had been now two months suppressed, I immediately printed a supplement, congratulating the Greek people on an event which the *British Star* had predicted two years before. I had about that time, in a series of articles, warned His Majesty of Greece of the approaching danger, and pointed out the means of averting it.

In the supplement which formed the final number of the last volume of the *British Star*, I appealed to the good sense of the Greek people, and advised them to elect Prince Alfred to be their king. I pointed out the many advantages that must accrue from such an election, and amongst others, the inevitable annexation of the Ionian Isles to Greece. Until the appearance of that supplement no individual, no pamphlet, no journal, except the *British Star*, had ever pronounced the name of Prince Alfred in connexion with the throne of Greece. Two years before the dethronement of King Otho, the *British Star* had openly advocated the nomination of the English Prince as his successor. There had also appeared, in the columns of that journal, an account of the organization of the French National Guard, recommending that that of Greece should be constructed on the same model. Indeed, I printed this

o

essay in a separate little volume, and distributed it, *gratis* among the Greeks.

King Otho being removed, my supplement was able to circulate freely through Greece. It fell like a spark amongst combustible materials. The flames spread in every direction. I distributed thousands of copies through the principal towns of Greece and Turkey, and I learned with unmixed joy that the general public, as well as the majority of the Greek press, echoed my ideas. True, I knew not how Her Majesty of England and the Prince might view the affair, but I was enthusiastic enough to believe that, if Prince Alfred accepted the offer made by Greece, the consent of his Royal parents would not be refused. As to the objections of Austria and Russia, these, I knew, could be easily overcome. The important point was to win round public opinion in England to favour the project. That was the moment for the Greeks of England to bestir themselves. Then they ought to have come forward and opened their purses. An exceptional case demanded an exceptional sacrifice.

But, unfortunately, the Greeks in London were divided amongst themselves. It is true that all were anxious to see Prince Alfred their king, but each wished to work according to his own views, and these views being modelled upon some personal interest, you may imagine how little chance there was of unity of action. Some well-known Othonites suddenly became Alfredites, and, stepping into the political arena, seemed determined to bear away the laurels justly due to those who had laboured to bring about the revolution. Some worthy merchants, who had remained in a state of political coma whilst the great battle was being fought, now threw off their lethargy, and, followed by their dependent brokers and clerks, suddenly became thorough-going demagogues. We every day see men who change their political opinions, but, in the present instance, these good merchants did great mischief. Being men of wealth, an unmerited weight was attached to their opinions. Their adherents and upholders confounded commercial capacity with diplomatic ability; they fancied that the successful merchant must be a wise politician. It must be confessed that the influence of mercantile habits strongly tinged the political proceedings of these newly awakened patriots. Though liberal even to prodigality in the expenditure of their words, they were wonderfully parsimonious in the bestowal of their pounds. Perhaps they wished to counterbalance a *flux de bouche* by an astringency of purse. Be this as it may, certain it is that committees were formed in which words were superabundant and subscriptions very scanty. The confusion was terrible. Every man wished to be leader; every man wished to be chairman; every man wished to appear in the English press as the head of the movement. Apprehending this internecine warfare, I had, before the appointment of the committees, proposed that certain distinguished English Philhellenes should be invited to become members, and place themselves at our head. I named several members of Parliament and other distinguished men, who I knew would willingly associate themselves in our enterprise, and who, both with tongue and pen, would support the elected of the Greek nation. One Greek, distinguished by his Russian sympathies, was afraid of offending the Czar, and declared that, to associate Englishmen in the movement, would be to introduce incendiaries into our homes. The Emperor of Russia would not recognize our revolution, he said; and to these objections the Greco-Odessian added many others equally well founded. This selfish narrow-mindedness provoked me. We Greeks were seeking to make an English prince our king, and were we to tremble before the Emperor of Russia? This was not the only occasion on which tempers were lost. The frequent conflict of Greek with Greek resulted in some angry shocks.

I might have passed over without comment the verbose nonsense with which so many strove to cover their mean ambitions and petty rivalries had these men come forward with the money needed to meet the emergency. But no; I blush to say that, though the Greeks of London possessed a great capital, to the amount of many millions sterling, with credit three times greater, these prating patriots kept their purse-strings closed. But, whilst so niggardly of their money, they were most liberal of their advice. By every post they sent voluminous despatches to the Greek Government, in which they never failed to recommend to their fellow-countrymen unity of sentiment and action.

Every day the Baltic Coffee House presented a scene of declamation and discussion that too frequently degenerated into downright quarrelling. Unfortunately for me, I was, on these occasions, brought into collision with the heads of many Greek firms with whom I had long been connected in business. These gentlemen now seemed to think they might dictate to me; they almost assumed a right to coerce my political conduct. All this was very worrying, the more especially as nearly every mail brought me letters from members of the Provisional Government in Greece, and from many distinguished Hellenes, urging me on the question of Prince Alfred's acceptance of the Greek crown, as if the question were in my hands. Disgusted and disappointed, I withdrew from the Greek Committees of London, where private ambitions and personal jealousies stifled the development of a healthy patriotic feeling.

I resolved to establish a Philhellenic Committee with the assistance of English Philhellenes, and with the aid of a well-known Member of Parliament I immediately set about the task. In order to give to this committee an official and independent character, I sent out to Athens Captain Nicolady, the ex-subeditor of the defunct *British Star*. The mission of this gentleman was to explain to the Provisional Government the object contemplated by the Philhellenic Committee, and to obtain an official authority for the execution of the design.

The anxiety I had so long experienced on account of my commercial affairs, combined with the annoyance and irritation attendant on the perpetual thwarting of my political projects, began to tell upon my health. The irregular hours at which I dined, long fasting, and eating when my stomach was quite exhausted, had disarranged my digestion. There were moments when I was so weak as to be scarcely able to breathe after ascending a flight of stairs. All these disturbances, mental and physical, at length culminated; the overworked nerves could bear no more, and, a few days after Captain Nicolady's departure for Athens, I fell dangerously ill.

CHAPTER XXXII.

MY ILLNESS.

I WOULD willingly avoid any allusion to my illness, as the recollection of what I suffered is most unpleasant; but much of what I have yet to relate in connexion with the Greek and Oriental Steam Navigation Company, the Philhellenic Committee, the Stock Exchange, and the Anglo-Greek Steam Navigation Company, hinges on the fact of my illness.

It was, I think, on the evening of the 22nd of November, 1862, that I returned exhausted and worn out, to Petersham Lodge. I had been very busy all day, and had not tasted food for nine hours. No dinner was prepared—the cook, through some misunderstanding, believing that I intended to dine at my club. Having to wait some time before dinner could be prepared, I sat down to play dominoes with my brother Francis. I felt every moment more and more exhausted. Suddenly the thought shot through my brain that I was unprepared to meet death. I started up, and said to my brother:

"Oh, Francis! I have not made my will. If I die, in what a state do I leave my affairs!"

I took a turn across the room, stopped short, and said:

"Oh, Francis! I think I am going to die. My breath is failing—I choke!"

All my blood seemed suddenly to rush to my head. I felt as though my eyes and mouth were filled with it—as though it had lodged in my throat and brain. My knees trembled, and a fear of death for the first time in my life took possession of me. I felt the veriest coward on earth. I took a glass of sherry. I had not tasted anything since nine in the morning; it was then eight in the evening. Leaning on my brother's arm, I walked out on to the lawn. The dark shadows of the large trees conjured up gloomy phantoms before my imagination. Terror filled my soul. Believing that I had only a few minutes to live, I hastily re-entered the house, went into the library, and, opening an iron safe, took therefrom a bundle of receipts. On a piece of paper I wrote my will in a few words, and, handing it with the receipts to my brother, desired him to give both to my wife on her return from town. I undressed, and went to bed. Within five minutes I felt internally a consuming heat, though my hands and feet were cold as ice. I thought death now, indeed, was about to seize me. I sprang out of bed, fancying that whilst I stood upright he could not lay his clammy claw upon me. I rushed out on the high road. The servants became alarmed; they brought me a basin of mutton broth, part of which I swallowed. After remaining a few minutes in the open air, my stomach rejected the broth. I returned to the house, seeking in the broad blaze of the gaslight a means of dissipating my increasing sense of terror. Now that my excitement subsided, I became drowsy and weak. I again went to bed. A pain resembling the lumbago, with which I had been some time before annoyed, shot round my waist, and seemed literally to cut me in two. I could not turn on either side, and at last, worn out by pain and excitement, I gradually fainted away. It was in this condition that

Dr. Julius, of Richmond, who had been sent for, found me. He quickly brought me back, and when, with the elaborate minuteness of a nervous patient, I had described all my symptoms—the rushes of blood, palpitations of the heart, &c.—he assured me positively that I had no organic disease; that none of the symptoms I described need give grounds for alarm. He said that over work and long fasting had disarranged my digestion; in short, he told me that I was suffering from a nervous attack. Never did drowning man cling with greater tenacity to a floating plank than did I to the presence of Dr. Julius. I fancied that, as long as he was in the room death could not touch me. Reassured by the doctor's explanation, I became gradually calm.

On the following day my worthy physician told me that I must not expect to recover at once, that I should be tormented for some time by these strange feelings. He added, that I must eschew all work, keep away from the City, and follow a strict regimen until my health should recover its former tone. I obeyed, child-like, and after a few days I was so much better that I believed I was quite well again. But the terrible feeling, which I call "thanatophobia," but which my physicians called nervousness, again overmastered me, and unfitted me for business during nearly two years. I was quite hypochondriacal. I could not see a large dog without fancying it was suffering from hydrophobia, and would infect me. Such fancies often crowded on my mind, but suddenly throwing them off, I would burst out laughing, and try to discover the cause of these illusions.

But the worst time was during the long hours that I lay awake at night, suffering from palpitation of the heart and twitchings of the facial muscles. I sometimes commenced the night by falling into a quiet sleep, and dreaming pleasant dreams. Suddenly horrible visions would rise before me, and I awoke in terror. I longed for daylight. How often have I, with fingers on wrist and eyes fixed on the chronometer, counted the beatings of my pulse. Were the pulsations regular, which was rarely the case, I became tranquil; were they rapid and irregular, as they were most likely to be, my fears rose to a painful pitch. I experienced great lassitude, want of appetite, and burning thirst. You will possibly smile to hear that I never rose of a morning without fancying that I had become the subject of some fresh malady during the night. The disease I most dreaded was consumption; and yet sixteen years before, when I really suffered from hæmoptosis, no such apprehension crossed my mind. Now, if my tooth-brush pressed a drop of blood from my gums, I immediately believed that I discovered in myself all the symptoms of phthisis. At another time it was moral fears that shook my mind. I began to fancy that my property was in danger, and I contemplated with horror the fate of my wife and children, and of all who depended upon me. I could not divert my thoughts by reading. When I looked at a book or journal a horrible amaurosis overspread my vision—I could not discern a letter. I abandoned my attempt to read in despair, and returned to the gloomy occupation of self-anatomy. I satisfied myself that I had evidence of organic mischief in all the great organs of my frame. At one time cancer of the stomach assumed the ascendancy in my imagination. Five physicians— Dr. Julius, Dr. Pettigrew, Dr. Fuller, Dr. Oldham, of Brighton, and Dr. Lovegrove, of Kensington—over threw that monster fancy. Then it was disease of the heart that rose before my mind. The five gentlemen who had refuted my hypothesis of the stomach disorder now tranquillized me about my heart. But no sooner had the doctors retired than all my worst suspicions returned. I *was* suffering from disease of the heart—why else these horrible palpitations?—but the doctors, for fear of aggravating the symptoms, conspired to hide the truth from me. I finally came to the conclusion that I was suffering from a complication of diseases, and that every organ of my system was attacked. Two or three times I fancied I was going mad—that I was really losing my reason—but the ridicule with which my friends assailed me obliged me quickly to abandon that fallacy.

I hope I shall never again pass such days as those of 1863. A hundred times in the four-and-twenty hours I went to the glass. I examined my tongue; I observed the expression of my countenance. I distinguished the *facies hippocrates*—the stamp of death; and yet my hair had not turned grey, nor had I lost flesh. Still I was changed. I looked careworn and anxious, which gave me the appearance of being years older than I really was. I no longer felt the cerebral vigour, the play of thought that prevails when Mind holds the supremacy over our being. I had a secret conviction that I should never again be what I had been two years before. To add to these miseries, I had now intermitting attacks of nettle-rash. What torment I endured! I fancied that I was covered over with what in Greek we call *niotis*. I could not keep a limb quiet; my hands and feet were in perpetual motion. I could not remain for five consecutive minutes sitting in the same place. Even the weight of my clothes was intolerable. I undressed, went to bed, and rolled in the cool sheets, hoping to allay the irritation of the epidermis. But over and above all the physical and mental torments I have enumerated, were the doubts that thronged my mind, shaking my religious faith. You will say I was hypochondriacal; I was worse—I was on the verge of misanthropy.

During the first four-and-twenty days there was little improvement in my condition,. though during that time I swallowed more than four-and-twenty bottles of medicine. At length I began to recover my physical strength, but my nerves were still unstrung. Terrifying apprehensions and painful fancies still floated through my brain. An appetite for food returned, but it was not a healthy desire for nutriment; it was a fever of the stomach, exhibited in paroxysms of *gourmandise* that impelled me to take more food than I could digest. The fears and fancies then became stronger than ever.

Meanwhile the Greek and Oriental Steam Navigation Company remained entirely in the hands of Carnegie, who, as I have already said, had arrived in London a few weeks before from Galatz, invited by me to assist me in the office. I was now obliged, for the first time, to give notice to my bankers that they were to honour all cheques and bills drawn by Mr. Carr. This gentleman, unfortunately, could not attend personally to the business, so the affairs of the firm fell into the hands of Carnegie.

Captain Nicolaidy, after an absence of three weeks, now returned from Athens. He was the bearer to me of the following diploma of the Provisional Government, to form the Philhellenic Committee, as well as an offer, from the Minister for Foreign Affairs, of the post of Consul-General for Greece in London. The state of my health obliged me to refuse the appointment, but my greatest mortification arose from the conviction that I was also too feeble to carry into execution the project of aiding my country as I proposed.

[*Translation.*]

Extract from the Report of the Meeting of the Members of the Provisional Government and the Ministers of State, held on the $\frac{22nd\ November,}{4th\ December,}$ *1862.*

"It has been decided to empower Mr. Stefanos Xenos, residing in London, to aid in the formation of a Philhellenic Committee, of which he has spoken in his letter to the Government, appealing for the successful accomplishment of this undertaking to the Philhellenic sentiments of the English nation, and to the patriotism of our compatriots in contributing to the moral and political advancement of Greece."

The Minister of Foreign Affairs has been requested to communicate this decision on the part of the Government to Mr. Stefanos Xenos, who, under all circumstances, has displayed the noblest and most patriotic sentiments.

The Provisional Government.
D. G. BULGARIS, President,
C. CANARIS,
M. ROUFFOS.

The Ministers of State.

TH. ZAIMIS,	A. DIAMANTOPOULOS,
T. MANGHINAS,	A. COUMOUNDOUROS,
D. CALLIFRONAS,	EP. DELIGEORIS,
D. MAVROMICHALIS,	B. L. NICOLOPOULOS.

Secretary to the Provisional Government,
  N. Chatzopoulos.

This is a correct copy.

Athens, $\frac{23rd\ November,}{5th\ December,}$ 1862.

The Minister of State for the Department of Foreign Affairs,
    (Signed)   A. DIAMANTOPOULOS.

If ever Greece needed the support of Philhellenes and of her children in England, it was at that moment. The voice of public opinion in this country was beginning to be heard in her favour, but the civil dissensions that had broken out in Athens threatened to throw a darksome shadow over the fair aspect of the bloodless revolution that had dethroned King Otho. Some daring demagogues in Greece thought to profit by the opportunity, and, in the absence of a sovereign, put themselves at the head of affairs. Everything at that moment depended on the Philhellenes and the Greeks of London. It was they only who could provide money, negotiate the establishment of the national credit of Greece, and upon them, in a great measure, depended the election of a king.

It would be difficult to express my vexation, my grief at finding myself incapacitated, at such a crisis, from serving my country. The annoyance I felt increased and prolonged my illness; in fact, I thought more about the prospects of Greece than I did about my personal and pecuniary interests, though I knew these were being undermined by the people associated in my business.

As soon as I recovered a little strength, I removed from Petersham to Brighton, but, far from improving, my health became every day worse. Spite of the assurances of Dr. Julius and other physicians, that dyspepsia was the sole cause of my illness, I still believed that I was suffering from some organic disease, which these gentlemen, through consideration for my feelings, would not avow. I did not follow the regimen laid down by my physicians, so that if I was two days well I was a week ill. Then came a fit of hypochondriasm. I sent for my doctors. They told me I ate too fast, that my food was not properly masticated, that it was not sufficiently mixed with saliva, and that too much labour was imposed on my stomach. My position was pitiable. I did little else than study my symptons. I had pains travelling from my throat to my shoulders, along my spine down to my heels, through my heart to the extremities of my fingers. I more than once thought I was about to be struck with paralysis or apoplexy. At length, after six weeks' inaction, I came to the conclusion that a life of idleness, where my attention was constantly turned to my symptoms, was the very worst I could lead. I therefore rallied my strength, determined to return to business, and take the reins of the company again into my own hands.

CHAPTER XXXIII.

RELEASE OF PETERSHAM LODGE.

On my return to the office, I found a great change in the aspect of affairs. Carnegie had assumed an attitude towards the representatives of Messrs. Overend, Gurney, and Co., as if he felt secure of succeeding me. My reappearance at the office disappointed, and the determination of my manner awed him. He wished me most amiably to go back to my house, and take care of myself, and not try my constitution so soon.

About the middle of January, 1863, Mr. Carr informed me that Messrs. Overends' book-keeper having struck a balance-sheet for the past year, found that we were losing, instead of making money. I replied:

"I knew that all along. Did I not write and warn them? Did I not tell you that when they disposed of my grain and steamers in their fashion? Did I not foretell the result? They drew accommodation bills on this grain for their own convenience. Now, what profits could we expect from such business? Did I not say to you, Mr. Carr, three months ago, that £50,000 would be wasted in mismanagement, misappropriation, and, in fact, downright robbery, by all concerned in that grain business? You may now try to throw the responsibility on me, and put the losses down to my account. But I will never consent. I will never agree to that. Messrs. Overend, Gurney, and Co. may pass in their books what figures they please, but I give you fair notice, Mr. Carr, that this time I will not bear such injustice. I would twenty times rather blow up the company, and bring everything before the public. I will not go on in this way any longer."

The boldness of my language alarmed Mr. Carr, who, indeed, had always acted most honourably both towards Messrs. Overend, Gurney, and Co. and myself. He could not deny that the losses sustained in the business were entirely the result of the sacrifices made of the grain. I do not know whether Mr. Carr reported what I had uttered to Messrs. Overend, Gurney, and Co. I only know this, that the pride of the giant capitalists revolted. They could not believe that any mere mortal—more especially one in feeble health—would dare to take such a tone with them. They immediately resolved to do with me as they had done with others: they would crush me by their commercial weight in the Bankruptcy Court.

It was a fortunate thing for me that, a few months previous to my illness, I had obtained from Messrs. Overend, Gurney, and Co. a letter in which they pledged themselves never to make any claim on my estate, Petersham Lodge. I had bought that property out of my own private resources, with the exception of £1700 advanced by Messrs. Overend, Gurney, and Co., and which I had repaid. For that letter, which now afforded me so much comfort, I was mainly indebted to the zeal and good feeling of Mr. Carr, who had strongly urged the great money-dealers to sign it. (See Appendix, Nos. 10, 12, and 13; but I had also to sign their tyrannical letter, No.11.)

A few days after my candid expression of opinion, the accounts were made out, when Messrs. Overend, Gurney, and Co. brought me in their debtor to the trifling amount of £380,000. Out of this, £280,000 might be set down to interest, commission, bonuses, losses incurred by others, misappropriations, &c. Against this amount they held my fleet and freights. These honourable Quakers insisted on my recognizing this account, adopting and signing it in every shape and form they pleased.

In the suffering and nervous state in which I then was, surrounded by political enemies and commercial opponents, you may imagine how all this shook me. But what made my position still worse was, that Theologos, of Galatz, to whom I had opened credit to buy nearly 20,000 quarters of wheat and Indian corn, and who was bound to send the bills of lading, after shipment, direct to me, profiting by the complications that had arisen, forwarded some bills of lading to Carnegie, in London, and delayed the shipment of the rest of the grain, waiting to see from what point the wind was likely to blow. When I asked Carnegie what had become of the bills of lading of vessels that I knew had arrived in the Danube, he replied, "The vessels have not yet arrived at Galatz to take their cargoes."

Truth to tell, I was glad to part with the Overends upon any terms. My physical weakness made the combat with these commercial giants too unequal. My rights, against a house of colossal credit, of overwhelming commercial influence—against a house whose trade policy was the most unscrupulous—had no chance.

CHAPTER XXXIV.

THE INSOLVENCY OF MESSRS. OVEREND, GURNEY, AND CO.

My friends in Greece will now learn what prevented me forming the Philhellenic Committee when the Provisional Government requested me to do so. But let us return to the affairs of the Greek and Oriental Steam Navigation Company.

It was about the middle of February, 1863, that Carnegie—who was now Mr. Carr's confidant—told me that Messrs. Overend, Gurney, and Co. had intimated to the "old man" (Carr) that they intended to wind up the concern. With his usual assumption of knowing everything, Carnegie entered into details. I must here remark, *par parenthèse*, that up to that moment this gentleman had never made the personal acquaintance of any partner in the firm of Overend, Gurney, and Co.; I do not believe that he had ever spoken to one of them. That, however, did not prevent his undertaking to give me all sorts of information respecting my own affairs. He told me that there had been great commotion at the "Corner House" on my account. According to his story, David Ward Chapman, Henry Edmund Gurney, Birkbeck, and Edwards were all in a high state of excitement. Chapman, he said, was my best friend, Edwards my most bitter enemy, Gurney an unmanageable Orlando Furioso; and so he went on to great lengths in the same strain.

"It is very extraordinary," I said, "that you

should have received all this information before I heard anything of it. Pray who is your informant?"

"Carr; and he also told me that if you consent to retire from the company he will be able to keep it going. He says he would give you all the brokerage, which would be worth from £4000 to £5000 a year. Then Theologos and Carnegie, of Galatz, can give you £1000 a year more by their commission business, if you retire."

I had heard enough. I saw through the whole scheme at a glance. There was Carnegie, once my clerk, now aspiring to become my master. I was to be his broker forsooth! No, no, Mr. Carnegie; things had not yet come to that pass. And yet his face was all the while so meek, his voice so low, his tone so humble, his manner so hesitating, that a looker-on could scarcely have divined in him an accredited ambassador.

Had my nerves not been so unstrung, I might have listened coolly to this presumptuous and daring, though meek-mouthed proposition. I might have taken a diplomatic tone, and played a wily game against the actors in the *tragico-comedia* that I saw was about being put *en scène*. But my fluttering nerves could not bear such a strain. Yielding to the irritation I felt, I told Carnegie to return to Carr, and tell him that I had no objection to the winding up of the Greek and Oriental Steam Navigation Company.

"Yes," I said; "let it be. I cannot now hope to save the company, and I am not disposed to work all my life for other people. I have no intention of remaining an instrument in the hands of Messrs. Overend, Gurney, and Co. for drawing their accommodation bills. But the company must be wound up.

The steamers must be sold; they must pass into new hands. The clerks must be discharged, the offices closed, and no trace of the Greek and Oriental Steam Navigation Company left. To act otherwise would be to leave the company in existence for the benefit of you, and Carr, and the Overends, which is what I never will do—never. I would rather throw the concern into the Bankruptcy Court, and let the public judge the whole affair. The Overends are my sole creditors, or, more properly speaking, they are my partners, as I shall be able to prove. And I can tell you I esteem it no honour to be the partner of insolvents."

"Insolvents! nonsense," exclaimed Carnegie, bursting into a sarcastic laugh.

"Oh! do not laugh, sir; what I say is the truth. And if you knew the hundredth part of what I know, and what I have seen in that office, you would not laugh so heartily."

Carnegie became suddenly thoughtful. I saw what was passing in his mind. In shipping my private wheat, I was his debtor to about £2000; in shipping the company's grain, he had to take the acceptances of Messrs. Overend, Gurney, and Co. for from £30,000 to £40,000. He looked posed. But clear before his mental vision rose the colossal Overends. Who could believe aught against their credit? The shadow of doubt that had clouded Carnegie's brow passed away, my irritation cooled down, and, after some conversation about indifferent matters, he took his departure.

"The following day we met again, when Carnegie, in a jocular tone, said:

"Talking yesterday of the Overends, you said they were insolvent. What an idea! Their credit is so gigantic, their connexion so extensive, that, even if they are insolvent, a century may pass before they will stop payment."

"They will stop payment soon," I said: "yes, though they do not anticipate it themselves. At the next great crisis that shop will shut up, unless something extraordinary intervenes to supply their four or five millions deficit. You know that a firm may lose £4,000,000 in a few months, but that years may not repair the loss."

Carnegie twirled his moustache in evident anxiety. He was quite willing to become the champion of his future patrons, but he felt that the reins were not yet quite out of my hands. He knew that I was resolute of spirit, and he did not wish to incur any risk until he should be sure of his position. He questioned me minutely about the solvency of Messrs. Overend, Gurney, and Co.; and as I knew that what I said would be carried to their representative, Mr. Carr, and having no objection to their knowing that I was not ignorant of their position, I did not hesitate to state my views regarding the great monetary house.

"It is a great mistake," I said, "to compare great commercial houses with large or small kingdoms, as political-economists sometimes do. It is a great mistake to believe that, because a kingdom may, with some show of reason, be compared to a cat, that is said to have nine lives, and dies slowly, and sometimes revives at the very moment that it is expected to expire—it is a mistake, I say, to make the same comparison with regard to great commercial houses, which command vast credit though actually insolvent. Instead of dying a lingering death, such houses are suddenly submerged in a crisis—they fall with a crash, like some mighty tropical tree uprooted in a storm, striking to the earth those who sought shelter beneath its branches. But, laying metaphors aside, let us come to the stern facts of arithmetic. The original capital of the 'Corner House' could not be, when I knew them, more than £700,000 or £800,000. It is said that Robert Birkbeck brought to the firm £100,000, and Chapman's sons another £100,000. Now, it is rumoured that Mr. Barclay Chapman, the father, after the famous case of Davidson and Gordon, retired from the firm, or was compelled by the other partners to retire, in 1858, on the payment to him of £250,000 to £300,000. So, if we assume that the collected capital of all the partners amounted to £900,000, and deduct this £300,000, and £200,000, the dead loss by Gordon's case, we shall see that the capital, even in 1858, was reduced to £400,000 only.

"Now, in order to ascertain their real position since that period, we must, at starting, take into consideration two points—first, that, surrounded by rogues, and indulging personally in extravagance, the profits that appear on their books may be only imaginary; secondly, their annual public and private expenditure during the time when they were losing and not making money.

"Now with regard to the losses. Did they not, by Gordon's bankruptcy, lose £200,000? Did they not, by the Galway Company, lose from £700,000 to £800,000, it might have been even a million? They lost £100,000 by Zachariah Pearson, £300,000 more by the East India and London Shipping Company. Did they not lose £200,000 by Messrs. Lawrence and Mortimer, the leather people? They advanced on the Millwall Works, or Mare's yard, £300,000—works which you know are not worth £40,000. In the crisis of 1860 they lost £350,000 by bad bills of Lascaridi, Psichari, Baltazzi, Agelasto, Vlagomeno, Rodocanachi, and thirty other foreign, London, and Manchester houses. They lost £400,000 by Leopold Lewis's business. To all this let us add six years' office expenses. Let us take into consideration the enormous expenditure of the different partners of the firm, and the sums spent on legal advisers, favourites, and in other ways. For all these I set down only £80,000 per annum, or £480,000 for six years. I estimate the loss on imaginary securities at £100,000. The mode in which the affairs of the firm have been managed by the present partners, has given an opportunity to dishonest men to step in and make for themselves at least £100,000.

"But who can calculate the losses sustained by Messrs. Overend, Gurney, and Co. in their dealings with several great railway contractors, to whom they advanced hundreds of thousands on securities that cannot be realized, or yield a dividend before the completion of the works?

"From this rough calculation you must perceive that the great money-dealers have lost their capital, and must be insolvent to the amount of £3,000,000 by losses sustained through bankruptcies and moneys advanced on insolvent estates. They have sunk £3,000,000 in securities that will not ultimately fetch a third of their reputed value. Now let me ask you, whence comes the money now expended, but from the daily deposits made by the public in their bank? And is it not a fraud, a sin, a crime to employ the money of trusting people in that way? If it were not the depositors who supplied these millionaires with money, they would make accommodation bills with every man with whom they have pecuniary transactions. They would take his property and his acceptances, remortgage that property, and rediscount those acceptances with their indorsements. Look at this company, the Greek and Oriental. I shipped grain to make a profit, as you know I could do, but Messrs. Overend, Gurney, and Co. turned my grain business to an accommodation trade. They gave three months' acceptances, and no sooner did they get possession of the bills of lading, than they gave the grain to the factors, Messrs. Coventry and Sheppard, making them accept for a larger amount—viz., for the grain, freight, and insurance. This done, they abandoned the sale to their discretion.

"Take another instance. They advanced me money on my steamers, and they remortgaged for a larger amount these same steamers to Masterman's Bank, the Bank of London, and others, making me pay no end of interest, and giving an opportunity to others to profit by such transactions. Are these the acts of solvent capitalists, of upright merchants, of honest-minded men? Have I cause to tremble before these men, lest they should put me into the Bankruptcy Court? I owe nothing but what I owe them, with the exception of a few hundred pounds, which I can pay at any time. But, in point of fact, Messrs. Overend, Gurney, and Co. are as much my debtors as I am theirs. Have they not taken possession of the fruits of my six years' hard labour for these steamers? Have I not been most shamefully treated by their favourites, under their authority? Have I not been a loser by their faithless conduct, their illegitimate trading, by the heavy interest they extorted—in short, by that unprincipled system carried on under a broad-brimmed hypocrisy, a system under which many an honest merchant of the City of London has been surreptitiously crushed? No, no, Carnegie; that system will not be applied to me this time!"

I here paused for breath. I was exhausted from excitement. After a few minutes, I resumed:

"Has not my health been broken, have I not been brought nearly to the verge of the grave by these Overends? Had I been allowed honestly to clear my little fleet of steamers, I might now be in possession of a comfortable fortune with which I could retire to my native country. And what a position might I not occupy at Athens—what services might I not render to Greece! These men have broken me down physically and commercially, but I defy them to break me down morally. The lesson taught by the fate of Zachariah Pearson and others has not been lost on me. You think the Overends would go to extremities with me. I tell you they dare not. Should they attempt it, I am prepared to send out this circular at once."

Here I took out of my pocket a printed circular, in which I announced to my bankers and the public that the Greek and Oriental Steam Navigation Company was compelled to stop payment, because Messrs. Overend, Gurney, and Co., my partners and capitalists, could no longer supply the business with the necessary funds, being themselves insolvent. I appealed to the good sense of the public, when the details should be known, to decide upon whom the blame ought to rest. You must not forget that the Greek and Oriental Steam Navigation Company had no creditors worth speaking of except Messrs. Overend, Gurney, and Co. By the following letter it will be seen that Messrs.

Overend, Gurney, and Co. engaged to pay all the expenses of the steamers:—

The Greek and Oriental Steam Navigation Company,
19, London Street.

London, 26 July, 1862.

GENTLEMEN,

Referring to the supplementary mortgage executed by you this day in my favour, whereby you assign to me the freights of the ships therein mentioned, it is understood that in the event of my securing any of those freights, I am to pay thereout all accounts for insurance, stores, wages, &c., properly incurred for the voyage in respect of which such freight is payable.

I remain,
Yours truly,
E. W. EDWARDS.

Carnegie, upon hearing all this, and seeing the circular, stood fixed, as if rooted to the ground. It was not difficult to read in his face that he longed to flee to Mr. Carr, and tell what he had heard. I knew full well that he would report every word I had uttered to Mr. Carr, who would carry it to Messrs. Overend, Gurney, and Co. This conviction made me speak all the more frankly.

CHAPTER XXXV.

THE STORM.

It was about the time this conversation occurred that Messrs. Overend, Gurney, and Co. sent me a copy of the account current between them and the Greek and Oriental Steam Navigation Company. In this account my debt to them figured at the respectable amount of £383,039 8s. 2d. The document was accompanied with a polite request that I would acknowledge and sign it.

I wonder I did not drop down in a fit when I glanced at the enormous " total." As it was my sight became dim, and the figures danced before my eyes as I turned page after page of the terrible document. I doubt whether the commissary-general of a great army, in furnishing his accounts after a campaign, would cover more paper, or make a more terrible array of figures. Dazed and bewildered, and feeling as if half suffocated, I at length stood up, locked the ominous document in my private safe, took my hat and hurried into the street. I never stayed my steps until I reached London Bridge, and there, leaning against the parapet, with laboured respiration I tried to catch the fresh river breeze.

I said within myself: " What do they mean? Do they wish merely to frighten me, or do they want to drive me, like Pearson, into the Bankruptcy Court? I have not had £383,039 8s. 2d. of them. Surely my steamers, making as they did such profitable voyages, could not have entailed losses to the amount of £200,000. Surely Mr. Edwards, the Gurneys' great mathematician and accountant, their nominee and legal adviser, will not urge them to such a step."

Reasoning thus with my own spirit, under the play of the free river air, I gradually recovered my tranquillity of mind, and retraced my steps to my office, at No. 36, King William Street, determined, ill as I was, to fight my foes. Having reached it, I put the dreaded accounts into my pocket, and started by train for Petersham. By the time the signal whistle sounded I had concentrated my patience, and commenced to go item by item through the long debit and credit columns. The result of this examination was that in the evening I wrote the following letter * to Mr. H. E. Gurney, which on the morrow I sent to his private address:—

February , 1863.

H. EDMUND GURNEY, Esq.

SIR,
The Greek and Oriental Steam Navigation Company presents indeed a terrible state of affairs. I knew it, and I remonstrated long ago; but I was not only not listened to, but the management was taken entirely out of my hands. Third parties stepped in and made enormous sums of money, and confusion and discredit succeeded the organization that I, with pains, risk of life, and loss of health, had established. What is done now by others appointed by you, and *not* by me, is very difficult to cure; but I am astonished to hear that I am looked upon in your office as the principal cause, and that I am called by names I do not deserve.

To condemn, Sir, a person without a fair trial, is an injustice; and you, I am certain, Sir, will never do so. You appointed Mr. Edwards to superintend our finances. Mr. Edwards, having other engagements, never came to the office to see the accommodation

---

* I give here merely an abstract of the letter, to avoid the repetition of statements already made in the earlier pages of this work.

bills of Barker and Lascaridi. Lascaridi also deserted the company. I was left alone to manage 24 steamers. One night, to my great surprise, Mr. Edwards came and said that it was your wish I should sign a partnership agreement with Lascaridi. I objected to this, saying that my partners—if I had any—were Lascaridi and Co., and not G. P. Lascaridi; but Mr. Edwards insisted, and against my will, by *your order*, took my signature at the bank and the management of the finances. The result was that in three months there was a loss of £12,000 out of accommodation bills, without taking into consideration what you have paid for him, Manuel, and Valsami. I ask you, am I to be blamed for that? You did not trust me, you trusted him.

Mr. Barker, by your orders, succeeded Mr. Lascaridi to finance the company; you paid the money to him. What is to be said about that? Still I protected your interests. I saved you every farthing when it was time and I was advised by friends of yours to protect and provide for myself. I was master of the business—sole owner—and could easily have had £50,000; but I saved you that money, and not only that, but even my house is not settled, as I intended, on my children, but left to the current of the stream.

Mr. Henry Barker financed the company, charging several thousand pounds for interests and commissions on accommodation bills. At last you sent Mr. G. B. Carr into the office—a gentleman for whom I have great respect, and who served you with unparalleled faith and anxiety.

You will find in my letters that eighteen months ago I protested against the forced sale of my grain by your order; it was always sacrificed at 5s. to 10s. per quarter under the market, and I returned the contracts. There is a loss of £35,000 to £40,000 in two years. Shall I add that Mr. Edwards, to oblige Mr. O'Beirne, chartered, as mortgagee, in the affair of the Trent, the Palikari, and made the Admiralty refuse immediately the time-charter of the Palikari and the Mavrocordatos that I had arranged with Preston, and consequently I lost £10,000 net profit?

Without going into the details of the books, our debt to you stands now for £383,000, and you hold security, according to the accounts sent to you, for £201,400. The deficiency of £182,000 does not arise because the steamers do not make money, but because they are overcharged by an extraordinary accumulation of debts, viz. :—

£95,000 interest and commissions are charged in your credit.
30,000 to Barker Brothers, and others introduced through him.
10,000 bad debts of Lascaridi.
45,000 losses on the grain by forced sale.
15,000 law.
22,000 repairs and improvements of the steamers.
20,000 loss by the sale of my steamers to Pearson.
 2,000 loss out of sale of Colocotronis.
25,000 to receive of debtors.

£264,000

If the company had the capital from the beginning to build steamers for the trade, and had not to pay the above enormous amount of £264,000 for interest, commissions, and unavoided losses, the company would not be involved as it now is, but the debt of £383,000 would be only £119,000 against property of £201,000, leaving £82,000 profits from 1860 to 1862—viz., a sum above 10 per cent., without adding many other thousand pounds' profit that I lost by the interference of others.

This is the simplest way to put before you, Sir, our affairs. I did all you wanted of me. I signed every paper or letter you brought me. I recollect, in the Greek crisis, Mr. D. W. Chapman, in the presence of Mr. Edwards, requested me to sign the first letter you wrote consigning all the steamers to you. After I had done so, he said: "Now, without criminating the office, now that you have signed, I shall do something handsome for you." And now, after two years more labour and sincere work, the handsome thing is to be menaced with the Bankruptcy Court.

      I remain, Sir,
       Yours obediently,
        STEFANOS XENOS.

On the following day Mr. Carr was anxious that I should sign this account. I steadily refused; he persisted. High words ensued. I now, for the first time, repeated to him all that I had on the previous day said to Carnegie with regard to the great capitalists.

I observed that Mr. Carr's communications with the house of Overend, Gurney, and Co. were now more frequent than before, at the same time that he put his hand on every piece of property within his reach. For example, he went to the French Consul, and received £7000, for which he gave a receipt in my name. This was for the freight of the steamship Palikari, that had been chartered by the French Government to take troops to Mexico. Upon Mr. Carr's proceeding I made no remark, though I could have held the French Consul responsible for the money. I was resolved to keep on the safe side. I knew that if the Overends could take the slightest hold of me, they would not spare me. I understood their game. Dead men tell no tales.

Meanwhile I was urged in a friendly way to sign the account current. I persistently refused.

On the 27th of February Mr. Carr did all he could to persuade me to sign this account. He said the great capitalists had promised that no sooner should I sign than they would give me a release from all their claims. I laughed.

"Release me!" I said, "and from what? I am free now, because I have not recognized these accounts. If I sign, I bind myself—I throw myself into the power of Overend, Gurney, and Co. I should then have to look for a release to men who have broken faith with me so often. Oh, no; I'm not to be caught that way; I will not sign, and I do not fear them. If I become a bankrupt, they will be the cause. Overend, Gurney, and Co. are worse off themselves than I."

Mr. Carr laughed aloud. "You're mad," he said. And I believe he thought so at the moment.

"I am not," I replied. "I have their losses written down on a paper here. Let who will come forward now and tell me the Gurneys are not insolvent to the amount of £3,000,000 or £4,000,000.

Mr. Carr took up the paper, and, after glancing over it, said:

"If these are their losses, where are their profits? What do you know about the great profits of such a colossal house?"

"'Tis impossible there can be profits with their mode of doing business. And look at the troop of rogues collected round them. Now I ask you, was there ever a better business introduced into Lombard Street than the Greek and Oriental Steam Navigation Company? And what is my position to-day? If there was really a loss of £200,000 on my steamers, where did the money go to? I didn't get it. I wish I had even a part of it. It was wasted, poured out like water on sand, and no trace left. This is the result of their mode of doing business. Now, where are profits to come from? However, my feelings towards Mr. Gurney, as I have said, are the same as ever. He is the dupe and victim of Chapman, Edwards, and Barker. You know these men. You know that every consideration yields to the advantage derived from a paltry commission. They introduce every kind of rubbish into Lombard Street, and persuade the Gurneys it's all pure gold. How they do it I don't know. This I believe, that Edmund Gurney is an upright, honest man. At half-past four he quits the City, and goes home to his family, leaving his legal adviser and partners to look into the clauses of the agreements made during the day. Do they stop in the office to do this work? Not at all. They go on their way in another direction to amuse themselves, allowing the business to go to the dogs. But the day will come when the Gurneys will discover their real position; but it will be too late—too late. When Mr. Gurney transferred my steamers to Pearson —when he gave orders to Coventry and Sheppard to sell my grain—I could see clearly he was acting under advice. I felt he knew nothing of the real state of affairs. He was put forward and saddled with the responsibility to impose on Pearson and me. That was done by Chapman and Edwards. Now, Mr. Carr, you know I'm not stupid; a hint is enough for me. Give me a glimpse, and I can penetrate into a great depth. I can read the game of that great house perfectly well. There are the Chapmans and Edwards, who pull the strings that make the puppets dance; but there are the Gurneys and the public, who will be the victims. I wrote a letter to Edmund Gurney. I shall neither sign nor take any step until I get a reply. When that comes I shall know how to act."

Mr. Carr listened with his customary patience and kindness, though he was evidently anxious to fulfil the mission on which he was sent, and get me to sign the accounts.

Scarcely had Mr. Carr left my room when Carnegie entered. He was very desirous to know what the "old man" had said. What he said himself would not be, under ordinary circumstances, very gratifying. He informed me that, in the event of my retiring, he was to be appointed my successor by the great capitalists.

The idea of retiring from the Greek and Oriental Steam Navigation Company had become familiar to my thoughts. I was growing weary of a struggle that, if prolonged, could not fail to bring me to my grave. And it was a struggle for mere money, not a race in the career of a high and noble ambition. I knew the value of life too well to deem the prize equal to the sacrifice. Besides, was I certain of gaining the prize? After ascending hills innumerable, I saw before me the task of scaling mountain after mountain. No man knows the value of money better than I, and no man was ever more willing to work hard for it. But I regard the acquisition of money as a means, not an end; and I felt it would be a pitiful frustration of an end to sacrifice life in mastering the means that should lead thereto.

CHAPTER XXXVI.

A CHALLENGE.

THE question of signing the accounts one day brought on an unpleasant collision between Mr. Carr and me. The next day he began to threaten me, and wrote the letter No. 23 in the Appendix.

Mr. Gurney's reply to my letter arrived at length. It was verbal; Mr. Carr was the medium of communication. The great capitalist, I was informed, had read my letter attentively, and had spoken kindly of me, but was determined to wind up the Greek and Oriental Steam Navigation Company; it was therefore necessary I should sign the accounts. Having delivered this intelligence, Mr. Carr drew from his pocket a slip of paper, on which I saw some lines in the handwriting of Mr. Bois, the chief clerk at Overend's. Mr. Carr asked me to transcribe, with my own hand, at the foot of the account, what was written on this slip of paper, and then put my signature beneath. I have the scrap of paper still in my possession; it contains the following:—" Memorandum. The above accounts have been examined, and the various items checked and vouched, and such accounts have been finally balanced and agreed, and the same are adopted and acknowledged to be correct, and that the balance owing by it to Messrs. Overend, Gurney, and Co., on the 31st of Dec., 1862, amounts to £383,039 8s. 2d. London, 3rd March, 1863."

I leave the reader to guess who was the legal

adviser that drew up this little document. Having read it, I said: "Do you really think that I am such a muff as to sign what is untrue? I never checked or vouched or acknowledged these accounts, Mr. Carr; I therefore will not sign them."

The old man's face became livid, his lips quivered; he looked fiercely at me for a moment, then losing all command of his temper, he said in a loud voice:

"You're a blackguard."

"You call me a blackguard," I said, "because I refuse to acknowledge the robberies of others." Laying my hand on the old man's shoulder, for I had completely lost my temper, I ordered him to quit my office. Then going to the clerks' room, I said: "Let no one here obey him in future; I am sole master of this office."

Mr. Carr, I have no doubt, was sorry for what he had said. He silently left 36, King William Street, but as my clerks heard of this *fracas*, I was much dissatisfied, and determined at once to bring matters to a crisis. I took up my hat and followed Mr. Carr. I overtook him a few yards from my office.

"Now, sir," I said, "come with me to Lombard Street to settle this affair."

He made no reply, but accompanied me in profound silence to the "Corner House."

On our arrival Mr. David Ward Chapman was the only one of the firm visible.

"I want a few words with you, sir," I said, walking into the little room that looked into Birchin Lane.

Chapman observed my agitation, and, without making any remark, followed Mr. Carr and me. When he had entered the room, I closed the door, and said, looking him full in the face:

"I wish to know if you gave instructions to this gentleman to come to my office and insult me, calling me a blackguard because I refused to sign your accounts?"

Chapman looked interrogatively at Carr; the latter immediately said:

"I am sorry for having said so. I am ready to apologize to Mr. Xenos. I lost my temper."

"That's enough, Mr. Carr," I said; "not a word more. If you say you are sorry, I say I am sorry too —more sorry than you—that I treated you so. I say it sincerely. Here is my hand. And now," I added, looking Chapman full in the face, "I must tell you also, this, Mr. Chapman—whoever attacks my honour must be prepared to fight a duel unto the death. I have not suffered all this misery and trouble for nothing. You know well I have had in my hands hundreds of thousands of pounds. Thank your stars you had to do with Stefanos Xenos, who never took care of himself, who was fool enough to look always to your interests. And yet the money that passed through my hands was all earned by my own ships. Instead of thanking me for having saved you thousands, you now try to crush me with this account. And still, big as you have made it, it might have been £100,000 more, if I had looked to my own interests."

Chapman was silent and calm. I now confess I sought that interview for the purpose of bringing a protracted contest to a final close. I was sorry for what had occurred between Mr. Carr and myself. It was the first time we had ever had a difference. His advanced age and kind manners had always secured him respect. It was not in his nature to insult any one, and when he was rude to me I felt he was in a false position—sent to induce me to sign those accounts. Knowing all this, I went to Lombard Street to give a lesson to the proud David Ward Chapman. I threw down the gauntlet to him, but the ways of the "Corner House" were those of the Jesuits, not of the Templars. Chapman softly recommended me to sign the accounts, because *he intended afterwards to do something handsome for me.*

I replied that I intended to take advice, and would be guided thereby. I returned to my office, took all the books and papers connected with the Greek and Oriental Steam Navigation Company, and, having locked them in my private room, sealed the door. I then told the clerks they might amuse themselves until further orders.

During these movements Carnegie was acting the part of Mentor. He was trying to console the son of Ulysses, and cover him with the shield of Minerva.

Finally, I hastened to my old comforter, Hollams, and begged him to see Mr. Carr, with whom, before leaving the office in Lombard Street, I had arranged that he should meet Mr. Hollams.

Strange to say, I had all this time forgotten my bosom friend, Edwards. I said within myself: "We do not take a yacht as a gift, nor do we ride in the Park on a beautiful Arab horse, without remembering the donor. I shall write him a few words. Surely he has a heart. I shall test it."

I wrote a note, of which the following is a copy:—

MY DEAR MR. EDWARDS,

I thought, when I called on you the last time, that you would do something for me, and try to settle the unfortunate affairs of the Greek and Oriental Steam Navigation Company, particularly when you know that I am so ill; but, far to find some sympathy in your heart, as I expected, I am left in the most cruel position, for fifteen days, losing time and health. Now, dear Mr. Edwards, affairs arrived to this point that I must bring (one way or other) this matter to a terminus.

I left the City, I left my affairs to the hands of Mr. Hollams, who will call on you as my friend, and not as my solicitor, to see you and to explain to you my position. In this critical moment my life is more precious to me than all the property of the world; and in the way that I am suspended and tired it is the same as it is calculated to send me more to my grave than to the Bankruptcy Court.

Mr. Edwards never acknowledged my letter. He did not deign to reply.

To confess the truth, my spirit was now wound to the highest pitch. I was determined, at any cost, to save my honour from the machinations of the men into whose power I had fallen.

CHAPTER XXXVII.

THE SETTLEMENT.

THREE weeks elapsed, and I was still struggling with these men, invalid as I was, hoping to avert a public exposure of our differences. Harsh, ungracious, cruel as had been to me the conduct of Edwards and Chapman, I had no wish to exhibit myself before the eyes of the English and Eastern public, either as a commercial hero or a dupe, and there was no foretelling in which category I might be classed, or whether, like the old lady's mongrel pet, it might not be decided that I was "a little of each."

These were three weeks of hard struggle. Messrs. Overend, Gurney, and Co. could not believe that a man of shattered nerves, and actually suffering from physical illness, could resist the influences they were bringing to bear upon him. At this juncture I was more than ever struck by the zeal and integrity displayed by Mr. Carr in the service of such principals. I thought within myself: "The day is not far-distant when poor Mr. Carr's time and honest services will meet from these Pharisees the same recompense that mine did. When that day comes he will remember me."

During these weeks of struggle all my clerks came to know my position, and, with the exception of one or two, naturally sided with their expected future masters. I was not surprised. *L'argent fait tout.*

Meanwhile, I had frequent consultations with my

solicitor, who behaved to me more like a friend than a professional man. He had frequent meetings with Messrs. Overend's solicitors, and under his direction I signed the accounts between them and myself; not as they wished it, but " errors and omissions excepted." This I did to satisfy them, and in order to see if they really meant to come to a settlement.

Had this state of things continued three weeks longer I should certainly have been brought to my grave. The worst symptoms of my malady returned with redoubled force. I lost weight and flesh, I became pale, and was much troubled with amaurosis, double vision—in short, all those terrible sensations which I have summed up as "thanatophobia," again overmastered me. The very sight of the company's office doors in King William Street made me shudder. These doors seemed to me the portals of the infernal regions.

It was finally agreed that I should sell the Greek and Oriental Steam Navigation Company to Messrs. Overend, Gurney, and Co., as it stood, for £2500. They released me from all liabilities, and undertook to pay all the debts of the company, all the bills, and all just claims against the company already entered in the books with the knowledge of Mr. Carr. They were to take, through their representative, Mr. Carr, all my private wheat in London, and sell it with my concurrence. Out of the proceeds they undertook to pay my personal acceptances; should there be a deficit, it would be put to their account; if a surplus, it should be given to me. In short, they released me from all responsibility, and I agreed to assist them to settle the claims of the company and the pending accounts. They also agreed to give me eighty tons' space in one of the steamers, to transport my furniture to Greece, should I resolve upon going to that country. I was also to have accommodation for my horses, carriages, &c., &c. (See Appendix, Nos. 30, 31, 32, 33, 34, 35.)

Before I signed this contract, Carnegie was only too glad to give me the following letter, in which he declared that neither he nor his partner had any claim whatsoever on me:—

The Greek and Oriental Steam Navigation Company.

36, King William Street,
London, E.C., 24 March, 1863.

STEFANOS XENOS, ESQ.

SIR,

In consideration of your settling your affairs with Messrs. Overend, Gurney, and Co., and their relieving you from all the debts of the Greek and Oriental Steam Navigation Company, and to assist you in settlement, we do hereby also agree to relieve you from the balance of your personal debt in our books, and to look to the Greek and Oriental Steam Navigation Company for payment of same, and at same time you agree to do all in your power to cause us to be retained as the agents in the Danube of the said company.

Yours truly,
THEOLOGOS AND CARNEGIE,
of Galatz.

Who could believe that, within a few months after I had retired from the Greek and Oriental Steam Navigation Company, my two *quondam* clerks stepped out as proprietors of five small steamers trading in the Danube. The steamers were free of mortgage, and Messrs. Theologos and Carnegie were shippers of large quantities of grain. What was still more extraordinary, their *magazziniere* (wharfinger), Hadji Andreas, a bankrupt a short time before, when they took him into their service, now made his appearance as the owner of property worth £10,000. Galatz is a small town, and the proceedings of the new steam-owners excited universal remark and astonishment. A banker wrote to me of their goings on. But this is not all. Would you believe, reader, that my precious ex-clerks now began loudly to complain that I had caused them heavy losses; in fact, to the amount of £2000? Nay, more. Carnegie, even after the above letter, commenced an action against me in London on account of a bill for £200 or £300, an acceptance of the company's which Mr. Carr refused to honour, and which he as indorser was compelled to pay. The action against me was only a menace, and was abandoned at the very moment we were going to trial.

My contract with Messrs. Overend, Gurney, and Co. was concluded and signed about the end of March, 1863, at the office of my solicitors, Mincing Lane Chambers. I received the £2500, and transferred to them the steamers and a few cargoes of coal; then I issued the following circular to the shippers:—

36, *King William Street*,
4th April, 1863.

SIR,

I have the honour to announce to you that, owing to ill health, I am compelled to resign the management of the Greek and Oriental Steam Navigation Company to my honourable and able friend, Mr. G. B. Carr.

In thanking you for your patronage in the past, I trust you will continue the same in the future to my successor, as I am certain that by him your interests will be as well cared for as they have hitherto been by me.

I have the honour to be, Sir,
Your obedient Servant,
STEFANOS XENOS.

CHAPTER XXXVIII.

IDLENESS.

I DO not know what you, reader, will say upon hearing that I sold that great company for the paltry sum of £2500. It was currently reported in London that I received twenty times that amount. I do not now mean to excuse myself, but, finding that I could not untie the Gordian knot, I tried to cut it as advantageously as possible. If you reflect upon the complicated circumstances by which I was surrounded— if you consider my shattered health—you will, I think, admit that I took the most prudent course. I confess that the Bankruptcy Court had much greater terrors for me than it could have for Messrs. Overend, Gurney, and Co. I knew full well that Mr. Edwards, the official assignee for that court, would prefer the interests of his employers to my rights. Had I been a man of straw, the Bankruptcy Court would have been for me a short and safe purgatory, in which I could have been purified from the residue of all my transactions with Messrs. Overend, Gurney, and Co. But I possessed a few thousand pounds and my estate free of any mortgage, and the great money-dealers were my sole creditors. I had all the *matériel* of the *British Star*, and nearly 6000 volumes of that illustrated journal, which could now find free admission into Greece. To place all this property at the mercy of the lawyers and accountants of the Bankruptcy Court, at a moment when I had just issued triumphant from my political struggle with King Otho, would have been sheer folly. To enter into a warfare in the law-courts with a firm which, though rotten at the core, still presented, through a system of credit, a colossal front to the world, would have been as great a delusion as that of the Knight of La Mancha, when with his lance he attacked the thirty windmills. On the other hand, victimized as I was, I felt a kind of gratitude to Mr. H. E. Gurney, for the support that he gave me, and I felt sorry for his position. I had really no ill-feeling against the house. I knew that the Gurneys were ill-advised and duped. I saw that they were enveloped in so dense a mist, that they could neither see things as they were, nor discern their own position clearly. I lamented seeing them in the power of Edwards. But my regrets were useless. I could not enter the Lombard Street palace, and tell the head of the firm what I thought and what I apprehended. My most earnest wish was to get out of the fire before I should be singed. I saw ruin approaching the great capitalists *à pas de géant*. I thought within myself what would be my position if, continuing my connexion with them, and accepting bills for hundreds of thousands, they should suddenly stop payment. And let me here parenthetically remark, that they would have stopped payment in 1864, had not the mania for public companies come to their aid. They turned several of their *rubbishes* into limited companies, and got some of the public money. The bills of lading for the corn shipped on board my steamers were all made out in my name. They were sent by my agents direct to me personally. I could do with them as I pleased. I could sell the cargoes and keep the money, and fight their accounts;

but I never took advantage of them. The freights also were paid to me. I handed them to Mr. Carr, and he passed them to Messrs. Overend, Gurney, and Co. I alone had the signature in our bank, and I did not give it to Mr. Carr until I fell ill and could not attend to business.

On my return to Petersham, after my last battle, I became so ill that I did not think it possible that I should ever recover. Collapsing suddenly into idleness after the excitement of a perhaps too active life, I became a prey to the worst symptoms of the spectral malady that pursued me. And then domestic calamities fell upon me. Twice I was on the point of starting for Greece, and taking with me my entire household; but domestic causes interposed, and stopped the execution of my design. That was to me a *coup de grâce*. My health was now so altered, the *vis vitæ* was so lowered, that even in the middle of June I wore two winter coats, and that without feeling warm. That gloomy malady which my physicians described as dyspepsia and nervousness, but which I characterized as "thanatophobia," tormented me more than ever. This fresh attack was accompanied by a new symptom. Instead of suffering from a want of appetite, I was now subject to a fever of the stomach that impelled me to eat at every moment. Indeed, I could not be said to eat—I devoured large quantities of food. I afterwards suffered from terrible fits of indigestion, during which I was abandoned to the most gloomy hypochondriacal visions. Sometimes, in that low state, painful recollections of the Greek and Oriental Steam Navigation Company stole over my mind. I thought of the different characters that had taken part in that tragedy. I sighed to think how I had been duped.

I shuddered when I reflected on the opportunities I had lost, and the sums of which I had been robbed.

All this time I was not without *soi-disant* friends, who came to visit me, and tell what the world said and thought of me. What rumours I heard! One said that Messrs. Overend, Gurney, and Co. had taken the company out of my hands because of my mismanagement. Another report was that I had left it a beggar without a sixpence in my pocket. A third rumour declared that I had come off with £100,000, and was the cleverest fellow in London to get so well out of the great money-dealers' net. Having exhausted the history of my commercial affairs, my political deeds and literary works were brought under discussion. The introduction of such questions was, on the part of some of my visitors, the result of thoughtlessness or want of perception; in others, it was owing to an idle curiosity; and in some others, to a mistaken, though friendly zeal. But whatever the promptings that urged my visitors to speak, the effect produced on my health was uniformly bad. I was too nervous to bear allusions to the causes that had reduced me to the state in which I was.

After the first shock had passed, I summoned a philosophic patience to my aid, and endeavoured to compose myself by reflecting that the rumours floated by the *mouches de cour*, and the other *courtisans*, to gratify their masters, must die out, and truth eventually prevail.

Messrs. Overend, Gurney, and Co. having removed all impediments to their accommodation bills, were loading no end of cargoes of grain from the Danube. This gave a lustre to the firm of Carnegie that dazzled himself as well as others.

Upon my leaving the Greek and Oriental Steam Navigation Company, Messrs. Overend thought it would be a clever *coup* to change the name. It was accordingly rebaptized as the "BLACK SEA AND LEVANT STEAM NAVIGATION COMPANY." Nothing could show more plainly how little they understood the materials with which they undertook to deal. "Greek and Oriental" was a designation replete with meaning to the ears of the real supporters of the line. It meant the "orthodox" Greeks—men in whose hands is nearly all the commerce of Greece and Turkey. "Black Sea and Levant" implies the Turco-Catholic Greeks, who form but a very small portion of the exporters and importers of the Levant.

CHAPTER XXXIX.

ANNOUNCING TO MESSRS. OVEREND, GURNEY, AND CO. THEIR INSOLVENCY.

SUCH was my position when those who were now managing the affairs of the Greek and Oriental Steam Navigation Company, seeing that I remained perfectly quiet, and that I had not left England, as I at first intended, commenced a series of annoyances, such as those so admirably described by Balzac in " *Les Petites Misères.*" My brother, who had been freight clerk, was dismissed, and they refused to pay him his salary. Clerks who had been discharged by me were reinstalled. A small account for coals, amounting to £10, was sent to General B——, a friend of mine, in a letter, in which it was insinuated in an ambiguous manner that I was no longer connected with the Greek and Oriental. This amount of £10 had been placed to my debit in my private account long before I sold the company. There were also several other paltry annoyances. I only mention these things here as they caused a correspondence between Messrs. Overend, Gurney, and Co. and me, in which I was forced to speak plainly to them regarding their insolvency.

On the 20th of May, 1863, two months after I left the Greek and Oriental, I received the letter No. 36 in the Appendix, from the new manager, in which claims were made on me for coals I had sent to the Piræus and Smyrna in 1858, a year before I made the acquaintance of Messrs. Overend, Gurney, and Co.;

indeed, before I had given my fleet the name of the Greek and Oriental Steam Navigation Company. This was too much. I immediately wrote him the following reply, a copy of which, accompanied by a few words of explanation, I sent the same day to Messrs. Overend, Gurney, and Co. :—

May 20th, 1863.

G. B. CARR, ESQ.

SIR,

I need scarcely say with what surprise I read your letter of yesterday. How can you say that Mr. Ross, by *my* instructions, passed to my private account the sum of £351, the price of 270 tons of coal at 26s. a ton? Do you mean, in the face of three witnesses, to tell me that you know nothing about those coals? Do you mean to deny that I did not tell you nearly two years ago that those coals were my property, and not the Greek and Oriental Steam Navigation Company's, having been sent by me to Athens long before the establishment of the company—viz., in 1857-58? Do you mean to say that you did not agree at once, without any other observation than that instead of passing the coals at 30s., the price I asked, I must do so at 26s. only? The letter we wrote to Mr. Lucas Ralli at that time was after the clear understanding between us that you had nothing to do with that coal.

As regards the £180 12s. 10d. paid to me by Mr. Clenzo when in London, I told it to Ross at the time, and it is not my fault that he did not book them in my account accordingly. Ross is no authority for me, and you know the reasons for which I discharged him. If I was in your place, for the sake of those concerned in the business, you would never have taken him back. Ross is the man that told me, for the first time, in the presence of my brother Aristides, that Messrs. Overend, Gurney, and Co. are insolvent. He told me that March, Mr. Edwards's clerk, told him so when they used to remain in our office at night, by order of Mr. Edwards, making up the books of the company; and added, to my brother, that I had been treated very unjustly by Messrs. Overend, Gurney, and Co.

From the time I quitted the company, you have done all in your power to annoy me; whilst I, on the contrary, have spoken in the highest terms of you and Messrs. Overend, Gurney, and Co.

You remember your words when first you came to my office—
"Your desire was to see me a rich man in a few years, and then, your mission fulfilled, you would leave the office." But what has taken place? I am compelled, two years after, to retire, and immediately you turn out of the company's employment my relations and friends. I never deceived Messrs. Overend, Gurney, and Co. All the year round I told them, and I told you, and wrote them, that we were losing money, and that there was a terrible confusion. You, on the contrary, were always telling them that we were making money? Time will speak for and vindicate me.

Pray why do you not pay Papa's bill of £260? Do you mean to say that you have no knowledge of it? Did I not mention to you in time that Powell's bill was unpaid? Was it not in the bill-book, and in the hand-writing of your faithful Ross? Did I not, according to your wish, write to Alexander Bell to deliver to you any surplus of my wheat for the purpose of covering this £3000—viz., Powell's bill for £2500, and the two bills for £400? Why do you not pay the rent of the wharf? And why do you instruct the landlord to bring an action against me? Does not the wharf belong to the Greek and Oriental Steam Navigation Company? Are not the beams of the Asia, and other things belonging to the steamers, there? Did we not agree together, at the time, to hire the wharf for the purpose of placing there the stores of the company? Why do you repudiate the contract for coals? Had we any coals in Malta at the time I contracted? Did not our agents write us, at the time, that they could not get coals in Malta under 35s., and Houseden answer in English not to buy any, as we had made a contract in London? Is not that letter in our English letter-book at least fifteen days before the final arrangements for my leaving you the company? I can see now clearly your object. I care nothing and fear nothing; I have the most sincere friendship and gratitude to Messrs. Overend, Gurney, and Co.; yet I hope the Almighty, even if I am left penniless, will prevent me from ever being forced to ask a favour from them. My accounts with them are closed for ever in this world. Apart from their wealth and commercial influence, I am as much a gentleman as any of them, and shall therefore not trouble myself any more about their good or bad opinion. I shall give them any assistance in my power. But, mark this, as long as I have a drop of blood in my veins, I will not allow my character or honour to be attacked by anybody.

I declare, before God and man, that what I have written in this letter is to the letter true; and I call on you, if you are an honest man, to disprove it. Scarcely had I left the office when all of you showed your true colours. You discharged my relations, leaving them unpaid, and refused to accept their bills.* Although my father, an aged man, broke his leg on board the Mavrocordatos, for the dispatch of the steamer at Constantinople, yet you refuse to settle his accounts. The captain of the Mavrocordatos will witness that.

Do you think that such things can be passed in silence? Public opinion, which you are seeking so blindly, will judge us one day all according to our merits.

Yours obediently,
STEFANOS XENOS.

Many months passed away, and still my health was not improving; whether owing to dyspepsia, or nervousness, or hypochondriasm, I seemed to have lost all interest in life. I seldom went to town; I was incapable of making any exertion. I constantly received letters from Greece, describing the gloomy aspect of affairs in the kingless country. On reading these epistles, I felt like the wounded warrior, lying on the battle-field, who sees the danger he cannot avert. Since Prince Alfred had declined the throne of Greece, my political ardour had become as cool as though I were not one of the prime movers of the revolution of 1862.

I one day received a letter from Athens. It was from Mr. D. Mavromichali, the War Minister. He

---

* Several of the agents of the Greek and Oriental, in the different ports of the Levant, were my relatives, who rendered great service, by their influence, to the company. Immediately I left the company the new managers took the agency out of their hands, refused to honour their drafts for the disbursements of the steamers, and gave the agency, to say the least, to incompetent persons—to men who spared the fair fame of no one to justify the conduct of their masters in London.

painted in the darkest colours the political condition of Greece. I flung aside the letter as I had done others, but I could not forget its contents. I began gradually to remember how, six months previously, I had been appointed by my Government to form a Philhellenic Committee, to aid my country in the trying crisis when she found herself without a king. I thought how I had neglected my mission, and lain idle at Petersham, irritated and fretful. I thought of all this, and the reproaches of my conscience became sharp and bitter. I felt that I ought to *do* something. If in the commencement of my illness my physical weakness had depressed my moral energy, now the *morale* reacted on the *physique,* and by a sudden mental effort I sprang again into action. I immediately went up to town, and had an interview with Mr. W———, a Member of Parliament, and distinguished Philhellene. I begged him to aid my efforts. Mr. W——— took the affair in hand, and principally owing to his exertion the Philhellenic Committee was soon organized. This event took place simultaneously with the election of Prince George of Denmark as King of the Greeks.

A committee of such a nature established in this country, and consisting of influential members of the commercial and political world, was capable of conferring large benefits upon a small nation still in an unsettled political state. But when the small nation should have attained a condition of political stability, when its institutions should be brought into healthy action under a constitutional king, then the benefits arising from the action of such a committee as I have described would assume a social character, and would consist in helping to develop the natural resources, and so improving the future of the nation towards' which its friendly feelings were directed. Such were the objects contemplated by the Philhellenic Committee. That the proposed ends were not attained is attributable to causes which I shall now explain.

CHAPTER XL.

THE PHILHELLENIC COMMITTEE.

THE very day it became known in England that King Otho had finally departed from Greece, the Greek bonds, so long buried in forgetfulness, began to rise, and the national credit of the little Hellenic Kingdom, so long a dead letter in England, showed symptoms of rapid recovery. The election of Prince Alfred by the Hellenes had made a most favourable impression on the British public, and lighted up the warmest hopes, especially when the intention of transferring the Ionian Isles to Greece became publicly known. Greece became again a popular favourite in Great Britain, and could the ignorant leaders in Greece have discerned, as did the Greeks in London, the true path to national progress and power, Greece, instead of relapsing into oblivion, would now enjoy a respectable position in the minds of the English people.

In the account I have given of my commercial life, I have not once mentioned the Stock Exchange, and for the very simple reason that I had nothing to do with that institution. I had heard of so many rich men being swallowed up in that Charybdis, that I had a pious horror of its shores, and had always resolutely resisted the temptations held out by different stockbrokers. I had never bought or sold stock of any description, nor invested money in bonds of any kind. I have said how, three days before the event was known to the English public, I had received intelligence of King Otho's expulsion from Greece. What would that information not have been worth to other City men! I never dreamed of turning it to any monetary use. As I have said, after the revolution of 1862, the Greek bonds rose rapidly; in fact, they became for a short time a favourite stock with the British public—a stock, too, not of mere speculation, but of solid investment. Thus the Greek national credit was not only rapidly regaining strength, but was becoming unquestionable. The English public, reasoning from two points of view, gave to Greek stock its just importance. The Greek merchants of England had, during a period of forty years, represented the national character in this country. They had passed through severe commercial storms with unblemished honour and integrity. With this example before their eyes, the English said: "Now that King Otho, the great incubus of the country is thrown off, why should the Hellenes not act as uprightly as their compatriots in this country? King Otho being gone, the national honour remains in the hands of the Greeks themselves. We may, therefore, look forward to a better future, particularly as Greece needs money to put her finances in order and develop her resources."

These reflections were well based, and, I repeat, were it not for the gross ignorance of those who happened at that time to be leaders in Greece—were it not for their petty quarrels and silly jealousies—the national credit might have been permanently resuscitated. I saw and felt all this, and even during my illness had, through the Greek press, tested the feelings of the Athenian public on the question. The result was disheartening. A new generation had sprung up in Greece, amongst whom the belief prevailed that Greece had never received any other assistance from England than a few worthless ships of war. Some there were influenced by an adverse policy; some by a spirit of opposition; and others by selfish motives, who rejected the idea of paying the disputed public debt. Those who did not believe in the existence of the debt laughed at my letters, thinking them a sarcastic joke; others wrote to me, saying that, if I persisted in publishing such opinions, I should lose the popularity I had acquired; and others said that I furnished my enemies with grounds for hinting that I speculated in these bonds. Spite of all this, I persevered, and, through the English and Greek press, addressed the peoples of both countries. I felt that the question of the Greek bonds was not, so to speak, matured; I knew that the majority of the Greeks were not aware of the light in which the matter was regarded in England. I was bitterly attacked in many of the Greek journals, but I still thought it a duty to show how the question was really working. Now, neither in Greece nor in England was there any political question involved in these bonds; it was, therefore, evident to the perception of any practical man that the British public must be ultimately the sufferers.

The Athenians now showed a disposition to consider gravely the great national question of the development of the resources of their country. Anxious to stimulate these good intentions into action, I published in the English press several letters on Greek finance and the resources of Greece. I also printed a history of the Greek bonds. I earnestly recommended the Greek Government to profit by the opportunity then presented of raising capital in this country.

What a perplexing position was mine at that moment! Money-makers thought I had left the house of Overend, Gurney, and Co., to open this new species of Mint. My enemies in Greece stigmatized my patriotic exertions as a selfish effort to promote my private interests, regardless of saddling poor Greece with a debt of £7,000,000.

I can now satisfactorily explain the reasons that induced me to engage in undertakings which, far from being profitable, were really ruinous to my private interests. I have already said that, King George being seated on the throne of Greece, the only advantage that the Philhellenic Committee could offer to the Hellenes would be to aid them in developing the resources of their country. Simultaneously with the establishment of the new dynasty, the committee applied to the Greek Government for two concessions, the carrying of which into effect would confer on Greece, socially, materially, and politically, serious benefits. One of the proposed undertakings was the construction of a railway from the Bay of Arta, through Acarnania and the southern part of Aitolia, across the Gulf of Corinth, running through Athens, to terminate at Port Raphti or at the Piræus. The other project was to canalize the Isthmus of Corinth, and build a large commercial town there.

This railway, in conjunction with another which was on the *tapis*, and which was to run from Avlona to Arta, would have shortened the voyage to Egypt by three days, besides offering to travellers the immense advantage of exchanging a sea voyage for a comfortable land passage through two beautiful countries of

Europe—Italy and Greece. By this new route the sea passage would be only from Brindisi to Avlona, which might be performed by a good boat in three hours, and from Port Raphti to Alexandria, a trip of thirty-five hours. And, had the line been prolonged, as was intended, to the south of the Morea—that is to say, to the Laconian Bay—the sea passage would then have been only thirty hours.

Had this line been constructed, it would indubitably have been taken advantage of by passengers to and from Egypt and India to Western Europe; and how great would not the advantages have been to Greece! But the question arose: Would the line pay, and what guarantee would the Greek Government give the English public for the investment of capital? There could be no doubt about the railway paying, because, as the line would shorten the journey to Egypt by three days, the Governments of Northern and Western Europe, that pay large subsidies to the great steam companies, would find it their interest to patronize the junction of the lines of Italy and Greece. Over and above the profits accruing from the local traffic in goods and passengers, the Greek Government consented to give 100,000 acres of good land as a guarantee.

Those who understand the geographical structure of Greece must at once perceive that the canalization of the Isthmus of Corinth would be an enterprise replete with profit to the country. Vessels coming from the Black Sea, bound for the Adriatic and the Mediterranean, would pass through this canal to avoid doubling Cape Matapan, and consequently Corinth would become a dépôt for the bale and grain goods of the Levant and Black Sea. This would infuse new life into a now torpid region, and give wealth and independence to many who now live at the Government expense. Simultaneously with the concession for the canalization of the Isthmus of Corinth, we asked for lands necessary for laying the foundations of a town in the locality, and also for facilities for the construction of docks required for the repairs and accommodation of vessels trading in the Mediterranean, and so much needed in those waters.

I must say that, frequent as were the changes of Ministry in Greece at that period, each Ministry approved these concessions; and though I often thought them slow in bringing the question before the Chambers, or in granting the guarantee of required arable land, I must acknowledge that all recognized the importance of the undertakings, and endeavoured to put the concessions in the form most likely to pass through the Chambers without opposition.

M. Coumoundouros—without exception the ablest financier of modern Greece—analyzed and approved our plans. He prepared the bills on the subject to be presented to the Greek Parliament, and were it not for the early close of the Chambers that session these two concessions would have been ratified, and would have afforded abundant occupation to the Philhellenic Committee, and prevented its dissolution.

These two concessions being granted by the Greek Government, subject to the approval of the Chambers, and being supported even by the most influential members of the Opposition, all that remained to secure their realization was to raise in England the necessary funds. This was about the middle of 1863—a period when reigned in London that mania for public companies that brought about the late commercial crisis.

Nothing could have been easier for me and my supporters, at such a time, than to find the required funds, had the public credit of Greece been placed upon a proper basis in England. And that credit depended solely and entirely upon the arrangements that would be made between the bondholders and the Greek Government. It was the conviction of this truth that made me discuss so freely in the press the question of the Greek bonds. But my motives were misrepresented by my enemies.

I felt how ticklish was the position in which I was placed; but during eighteen months I unremittingly laboured for these two objects—to get the concessions approved by the Chambers, and to induce the Greek Government to make such arrangements touching the debt of 1824-25, as would make the shares of the concessions marketable in England. Had these concessions been carried into effect, it would have been seen that they were no illusory speculation, for not alone would they have proved of European utility, but would have yielded a handsome profit to the private individuals engaged in the enterprises. When King George passed through London on his way to Greece, I had several meetings with His Majesty's adviser, Count Sponneck, on the subject of the Greek bonds. I was charged by M. Rouphos, the President of the Provincial Government of Greece, to speak with His Excellency on the subject. In vain did I and some of the members of the Philhellenic Committee try to persuade him to recognize them, and telegraph to Athens that His Majesty could not go to Athens till the national credit should be redressed. Then the people and the ministers, who were anxious for the arrival of the king and the union of Greece with the Ionian Islands, would, no doubt, approve his acts, and pass it in the Chamber.

My health was now visibly improving. External occupation had turned me away from the disagreeable one of self-contemplation. My nervous system was regaining tone, and I wished for something more stirring.

Months had passed, and the Philhellenic Committee had not succeeded, either privately or officially, in coming to any arrangement with the Greek Government touching the concessions. M. Coumoundouros had them in his portfolio, and was watching the opportunity to pass them up to 1865. (See his letter, Appendix, No. 38.) In vain we represented to the Greek ministers that the rage for public companies in England made the moment propitious, and, if allowed to pass, such an opportunity might never return. These gentlemen took the matter coolly, and told us that the reorganization of home affairs, disturbed by the late revolution, was of primary importance, and demanded their chief care. I had already expended several hundred pounds out of my private purse in the getting up of the Philhellenic Committee, and keeping it up for two years; but, seeing no probability of a satisfactory result, I determined to transfer the scene of my labours to the east side of Temple Bar, and again try my chances in the City.

CHAPTER XLI.

THE STOCK EXCHANGE.

THE *tali* or *tesseræ* of the Roman gambler, of which Horace says "*vetita legibus alea,*" was a blameless recreation compared with that noble institution of honest gain known amongst us as the Stock Exchange. Here could be seen, in 1864, thousands of English and foreign *aleators*, laden with gold, crossing that fatal Rubicon, and in exciting tones exclaiming: "*Jacta alea est*"—" the die is cast." And soon these thousands are seen flocking back, presenting as miserable a spectacle as though they had, in returning, swum through the burning waters of Phlegethon.

When, in 1864, I returned to the City, after fourteen months' absence, I found all the professional occupants of that mysterious area in a state of *aleatorial* intoxication. The spirit of communism reigned supreme. Profits were to be shared in common; so were losses. That is the sense in which I read "limited liability." In commerce, every man is a demagogue; but there is this difference between the political and the commercial demagogue, that the latter does not traffic in lofty abstractions: he always deals in tangible articles, even though it be only *waste paper*, which he tries to convert into pure gold for his own benefit. There are times when these commercial demagogues are wholly disregarded, and exercise no influence on the public mind. Then there come times and seasons when these men are in high vogue, and hold sway in the monetary world.

When I returned to the City, I found these gentlemen in high fashion. They had come out that season under the euphonious title of "promoters."

On the morning that I made my *entrée* into the City, the first person I met was M. M——. This gentleman is a Greek, remarkable for rotundity of figure and a baritone voice. M. M—— has some peculiarities of manner as well as of person. He cannot speak ten words to you without either putting his hand on your shoulder, or catching you by the button. At other times he pokes you with his fat finger, and as you instinctively retreat at each poke, and he follows up his advantage, you ultimately find yourself either pinned to the wall or standing in the gutter, at the mercy of cabs and omnibuses.

"What, Polypathis!" he exclaimed—he used to call me by that name—"where have you been hiding all this time? You have lost the most splendid opportunities. We are making thousands of sovereigns every day. Just look here. I am proud to say that I am able to balance on my shoulders a head full of brains."

Several of the passers-by, attracted by his loud voice and theatrical gesticulation, looked and smiled. Had they understood Greek, how much they would have been amused.

"How do you manage to make thousands of sovereigns so easily?" I asked, not a little astonished.

"Oh! by my eyes, 'tis the easiest thing in the world. I have made £6000 ἄψε-σβέσε between lighting my candle and putting it out again. Look here, you need only buy shares in any company in which you know that one of our great Greek merchants will become a director; they will surely go up £2 or £3 per share. Sell immediately, and you realize your profits."

"Has the fact of a Greek merchant sitting on a public board so great an influence on the English public?"

"Yes, I tell you—yes. Watch the shares of the companies where the Greeks went in, and you will see how firm they are. The English know that the Greeks are the best financiers and merchants. They have faith in them. So the shares are firm."

"I know nothing about shares, my friend. I never played on the Stock Exchange until very lately, and I have had enough of it. When Lee crossed the Potomac, I thought the Confederate cause so safe that I invested some money. I bought at £90 per bond. A few days after, the Confederates were defeated by Mead, and the Confederate Loan is now down to £62. That, you see, is a heavy loss. So I have come into the City to sell the scrip, *coûte qu'il coûte*, and return to my old trade—the steamers."

"Steamers! Nonsense! Who cares now for steamers? Take my advice. Buy shares in the Metropolitan and Provincial Bank, in the Imperial Bank, in the Mercantile and Exchange Bank, in the European Bank, and others, of which our compatriots are directors. If anything goes wrong, we shall be the first to know it. Buy with several brokers, and as many shares as you can. The shares are sure to go up —right up, I tell you—in a few days."

Friendship may exist, though rarely, in the Westend of London, unmixed with jealousy; jealousy, I well knew, might be found in the City in its simple essence, without the slightest admixture of friendship. I had known M—— a long time. I had often given him a lift on the road of life. I did not entertain the slightest doubt of his sincerity. He was the only man I have ever met in the City of London who would indicate to a friend the road in which he was himself making money.

"If you are so certain," I said, "I shall try. Let us go round to the stockbrokers you know. I am acquainted with only two—those that bought me the confounded Confederate bonds."

In less than an hour I had bought 1350 shares in the Metropolitan Provincial Bank, the Mercantile and Exchange Bank, the Imperial Bank, and others, in the directorate of which M—— had great faith. All these shares were at that time already at a heavy premium.

My friend was not looked upon amongst commercial men as an oracle, but I must say that his deep-toned voice prophesied in this instance as truly as ever did the most sonorous Dodonian oak. Within three days I had sold all my shares, realizing a profit of about £2500.

The proverb says, "A fool may possess wit, but not judgment." I candidly confess that I acted the part of a fool, but it was not for want of judgment. I knew that shrewd and clever men had been ruined on the Stock Exchange. I felt that I had made a great *coup*, and I recognized the hazard of making a second attempt. Besides, I was just then preparing to go to Athens, to urge the Government to bring before the Chambers the concessions I have mentioned, but becoming infected by the company-mania, that proved fatal to so many others, I remained in London, and shared the common fate.

My friend M——, of whose office I had become an *habitué*, assured me that the shares of the Warrant Bank were sure to rise to the very zenith by reason of the expansive force about to be communicated by some great political economists who were going to join the board. Like the benumbed fly that has recovered animation in the warm atmosphere of a well-lighted room, but will not be content without fluttering round the flame of the bright-burning lamp, where it is finally consumed; so I, not satisfied with my first success, bought nearly 1500 shares, not only in the companies pointed out by M——, but in the International Contract Company, the Land Credit Company of Ireland, the Commercial Bank Corporation, the Freehold and General Investment Company, the City Offices Company, the Financial Discount Company, and the Joint-Stock Discount Company, all of which were at a heavy premium.

It has often occurred during my life that I have erected lofty towers, and just at the moment that I was about to roof them in, remove the scaffolding, and exhibit my work, a catastrophe, originating in some trifling and scarcely perceptible cause, has destroyed the edifice and annihilated my fine prospects. But, in the present instance, such was not the case. It would appear as though, after the inaction induced by my long illness, I required excitement, for I jumped voluntarily into the fire. I can only blame myself. The shares, instead of flying up, rolled down. In less than three weeks I lost not only the £2500—the profits of my first *coup*—but, unfortunately, I lost my presence of mind. I became obstinate. I bought shares which, according to public opinion, were likely to go up; I sold those which, on the same authority, were likely to go down. I was disappointed in both cases. This time the rhetorical prophecies of my friend M—— were falsified, and public opinion was utterly wrong. Even information derived from high and well-informed sources betrayed my expectations. Here I found myself in the midst of a terrible cross-fire, and how to retreat I knew not.

I shall not trouble the reader with an account of my adventures in Throgmorton Street and Broad Street during the year that followed my misadventure. Those twelve months might furnish comical sketches of many a group of bewildered Greeks. Only fancy those clever Hellenes, who had so often walked victors along Mincing Lane, Mark Lane, Eastcheap, and every monetary locality in the City—fancy these clever fellows, I say, caught like rats in traps!

My friend M—— lost all his gaiety. He smoked cabbage-leaf cigars through economy. Caught like Milo in the cleft oak, he said in his grief:

"Μωρὲ παιδιά (foolish boys)! who could have expected it? This Stock Exchange is like the serpent: catch the tail, the head bites you; catch the head, the tail strikes you."

"I thought," I said one day, laughing, "that you carried a head full of brains on your shoulders."

"So it was thought a year ago. But don't make me blush. I am like the dog whose master gave him a basketful of live crabs to carry home. On the way one jumped out. The dog set down the basket and ran after it. When he returned, he found two others had gone. He pursued them; meanwhile, three or four more jumped out. He found he could not succeed in rebasketing the crabs, so he killed them all, and put an end to his struggles. Now, that is my case. I had scarcely made £100 in Alliance Bank shares, when I lost £300 in the Scottish and Universal, or the City Offices, or the Bank of Mexico. There is scarcely a kind of stock in which I have not lost money. It's no use talking. To tell you the truth, I believe the Greek merchants who joined those companies were the cause of all our troubles. Every time they got in they made us buy or sell shares; yes, I tell you it was only to fill their own pockets. Don't you see? Yes—I tell you, yes—it is so."

And each time that he called out " yes," he gave me a poke in the ribs with his finger. Then he added, in a lower tone:

"Now, my eyes, Polypathis! what we must do is to save our credit. Foolish boys! we must keep our position; we must not lose that. Foolish boys! I tell you, yes. It is a great thing in this great town to keep our credit, for then the money comes again. Go ahead. My eyes, Polypathis! go ahead. Sell off, and pitch this infernal Stock Exchange to perdition."

I should not have thought of retailing these remarks of M——, did I not regard them as the nucleus of certain important facts, in the following chapter, connected with the limited company mania—facts with which the initiated alone are yet acquainted.

CHAPTER XLII.

MESSRS. OVEREND, GURNEY, AND CO., A LIMITED COMPANY.

THE Greeks of England are a formidable body, both as regards their wealth and commercial credit. Their tendencies are naturally speculative, and their inclinations *aleatorial*. When, in 1862, limited companies first made their appearance, many Greeks, seeing the premium that the shares bore, invested freely in the stock, but only as a matter of speculation. They realized great profits. This drew the attention of other Greeks, who at the time were losing largely in the corn market by reason of the great influx of American wheat and Indian corn. We shortly saw half a dozen Greek gentlemen start as stockbrokers. It was the first time Greeks had tried that line. These gentlemen soon persuaded their countrymen, with the exception of a few of what we call the old school, to buy and sell shares in these new concerns. The success of this undertaking induced half a dozen fresh Greeks to appear in the arena in another capacity. These men were employed by English promoters as under-promoters, and received a sum varying from £200 to £500 for every Greek merchant whom they induced to become a director in the new companies. At that time you might buy or sell any kind of stock without the least fear of being ultimately saddled with it. I am, however, bound to say that the Greeks were all "bulls," and never thought of selling but to secure a profit. With the quickness of a plague, this *aleatorial* epidemic ran, not alone through the Greeks of London, but extended to their compatriots in Manchester and Liverpool. It even reached Constantinople and other parts of the Levant. The consequence was, that the shares of a new company were often all bought, without discrimination, by Greeks—on speculation, remember. In this way, many fraudulent companies were enabled to float. The Greeks bought the same stock simultaneously, not through a spirit of unity, but from the habit of imitation. The shares naturally went up, and the English public, thinking there must be something in the matter, bought also, and so the shares continued to rise. I will give an example. The shares of the London Bank of Scotland, Limited, were 10,000 in number, and six Greeks of my acquaintance, all men of small means, held in one fortnight's account more than 5500 shares, because three of our Greek merchants were about to join the board of the bank. An already broken-down *petit* merchant held no less than 900 of these shares. He had bought them at £3 premium, and might have realized £2000, but he refused to sell, expecting the shares to rise to £15 premium.

From this position of things, three results must inevitably accrue. In the first place, the Greek directors of the respective companies could only realize a profit in these transactions by victimizing their compatriots by inducing them to buy shares. Secondly, the public in general would be victimized by buying what the Greeks would sell, either to realize a profit or escape with the first loss; and, thirdly, the jobbers and brokers would be severely bitten by dealing in shares that must ultimately become unmarketable.

As long as money was cheap, and confidence prevailed in the market, bills of exchange were easily negotiated, and the differences in the rise and fall of shares paid punctually every fifteen days to the stockbrokers. No one dreamed of the impending dangers. We must remember that about that very time several foreign loans, involving many millions sterling, were brought before the public, and favourably received. In negotiating these loans, a class of promoters different to that connected with the companies was concerned. They were men of higher position and standing. Nevertheless, the causes above mentioned were then beginning to be felt, and a paralysis was extending through every branch of London commerce. It was particularly felt in the Mediterranean trade, which within our circle, in 1865, became nearly a dead letter. Colonials and metals were greatly depreciated. Copper bottoms went down £15 per ton, refined Dutch sugar fell £4 per ton, Rio coffee 5s. per hundredweight, and yet no one would touch them.

To complete the sum of dark forebodings that foreshadowed the great catastrophe that fell upon the City of London, the termination of the American War was fast approaching. This, though a blessing to humanity, exerted an inimical influence on the cotton markets of Liverpool, Manchester, Leeds, Bradford, and other towns.

Still, spite of all these menacing prospects, under the dark shadow of which the ignorant and unconscious public was contentedly trafficking in shares, the terrible catastrophe might have been averted, and the ruin of thousands spared, had it not been for a new commercial *tour de force* adopted and carried into effect by a set of unprincipled men. This manœuvre seemed to throw a kind of hallucination over the public mind. People no longer discriminated between valid and worthless shares, between solid and fictitious business. The actors were merchants and bankers, the heads of firms that had long enjoyed the reputation of wealth, but who in reality had been during long years insolvent. These men now resolved to turn their firms into limited companies, and so not alone escape the punishment due to their misdeeds, but realize a handsome profit. These clever gentlemen contrived to sell to the open-mouthed public several millions sterling of bad debts, for which they received in return many millions in hard cash.

Messrs. Overend, Gurney, and Co. had given advances sometimes more than double the value on the property pledged to them. What a medley the realities of that great pawn-office would display! There were papers representing steamers, building-yards, cotton plantations, foreign railways, foreign land improvements, discounting companies, and I know not what. But these securities were worthless, so much more than their value being advanced upon them. This being the case, Messrs. Overend, Gurney, and Co. resolved, before turning their own firm into a public company, to convert some of this rubbish into separate limited companies. As Messrs. Overend, Gurney, and Co. could not themselves appear in these transactions, they had recourse to their old instruments, Messrs. XX——, who now figured in the City as great promoters. Not that they entertained a feeling of regard for these men, who had rolled that rubbish into their possession—the brokers who had induced them to advance money on such securities, heedless of their interest as long as they received their brokerage or their commissions—no! but these fellows were the few remaining members of the Cabirian mysteries of the Lombard Street house, and therefore qualified for the business in hand. On the other hand, the honest promoters' conscience was lightened by the prospect of relieving his old employers of the burdens he had thrown upon them. It is true that, in doing them this service, he hurled all the rubbish with destructive force upon many poor families —upon widows and orphans—but we must remember that there was fresh brokerage and large profits on the promised shares, to say nothing of the promotion money and minor benefits.

I must here mention that, before the great Lombard Street shop was converted into a public bazaar—before its wondrous treasures, which the public rushed so eagerly to buy, were exposed for sale—before, as I say, these events took place, the elegant David Ward Chapman retired from the firm. Some accounted for his retirement by saying that he had considerably overdrawn his account; others said he was expelled because the Gurneys ascribed the heavy losses of the firm to the obstinacy with which he had patronized the official assignee, Edwards. Whatever was the cause, it was evident to me that the proud David Ward Chapman was now too glad to descend to the rank of a promoter, rather than be one of the partners of the old or the new firm of the Gurneys. In this capacity soon we find him humbly seeking shelter, like many other wrecked speculators, beneath the roofs of Mr. Stubbs' or Mr. Perry's Panteleimonian Temples.

CHAPTER XLIII.

BATTLE BETWEEN THE JEWS AND THE GREEKS.

DURING all these transmutations, Throgmorton Street exhibited encounters unexampled in the annals of history. Under Titus, the Romans and the Jews came into severe conflict; but now, for the first time, the Jews and Greeks had an opportunity of fairly measuring their strength in moral combat. The Jews of Turkey had always been regarded by the Greeks as an inferior race, whom they employed as porters or couriers. But the Jews of Western Europe are a totally different class of men in every respect to their co-religionists in the East. They had long held a high place in society, and several had distinguished themselves in a larger area than the narrow precincts of Mark Lane. Now, for the first time, these keen antagonists—the Greeks and Jews—met in sharp encounter in the confined passes of Throgmorton Street. The terms of combat were unequal. These passes had for years been held by the Jews. But now the Greeks, rushing in a strong current from Finsbury, Threadneedle Street, Mark Lane, and Mincing Lane, endeavoured to clear the passage of the Jews, seize the pillars of Hercules, and enter the Hesperidian gardens that lay beyond. But, as I have said, the Jews had held these passes for years; they knew both the strong and the weak points of the locality. They had fought during fifty years more than one battle on the spot, and had always come off victors. They now met the Greeks with cool determination, repulsed their attacks, frustrated their ambuscades, and finally succeeded in creating divisions in the enemy's camp. After sustaining heavy losses, the Greeks were compelled to retreat.

Losses were entailed on the oldest and shrewdest members of the Stock Exchange by that sharp, short contest. Calculating that the conflict would last for years, these old foxes supplied both parties with ammunition and provisions, dealing freely in every kind of share. The sudden cessation of warfare left them overstocked with articles greatly depreciated in value. It was similar to what had befallen to several contractors at Constantinople, when the Crimean War came suddenly to a close. Many of these men had in a few months realized vast sums of money, but, on the unexpected cessation of hostilities, they found themselves overburdened with commodities, the value of which had rapidly gone down. What did the jobbers of the *aleatorial* house do in their perplexity? They used every stratagem, they employed every *ruse* to get rid of the depreciated shares; and they did succeed in gulling the public, and exchanged their worthless shares for solid gold.

It has been said, "To keep out of a difficulty is prudent; but once in it, to bear thyself so that the enemy shall beware of thee, is valiant." It was thus that the members of the Stock Exchange acted. They did not cease to deal in the questionable shares. They never allowed the difficulties of their position to transpire; but they established amongst themselves certain rules. For example, if they bought at 6 premium the depreciated shares of a company, they sold at 9; or if they bought at par, they sold at 3 premium. This explains how so many shares figured at this time in the list at anomalous prices. For my part, I think that this skilfully covered commercial retreat of modern times deserves to rank side by side with the military retreat effected by our great compatriot, the warrior-historian, and his ten thousand, in ancient days.

During this commercial warfare I have described, stock-jobbers sold to the combatants on both sides shares in certain new companies which they did not hold, and which, when the time for delivery came, they could only obtain by buying at tremendous prices. I must here make a short digression. We are aware that the Committee of the Stock Exchange has the privilege of making laws, which, though not the laws of the land, have as much force in the commercial world as though they were, and have on the minds of juries the weight attached to usage. This committee had about that time issued two *ukases*. The first decreed that no company of which two-thirds of the capital was not subscribed, and the deposit and allotment money paid up, could be considered to have floated. The second *ukase* declared that, where a settlement was refused by the Stock Exchange, all contracts with such company should be deemed null and void. These two decrees, though in appearance very beneficial to the public, did much mischief ultimately. By the action of these *ukases*, many good concerns were prevented from floating, and much rubbish was confidingly bought by the unenlightened public.

In virtue of these two Stock Exchange decrees, many solid and honest companies were refused a settlement, because some members of that gambling-hall had been entrapped as "bears," and would sustain great losses were such settlement granted; and, under the influence of these *ukases*, the public were encouraged to take part in worthless enterprises, because the promoters of these undertakings had been clever enough to entangle some members of the great *aleatorial* mansion in their affairs, and so make them parties to their speculations on the public credulity.

Things were in this state when the first roll of the thunder was heard, and the electric current struck with fatal force the Bank of Leeds, an old and much-respected concern. This was a warning correctly interpreted by many, and might have been widely recognized but that a few of the promoters contrived to blind some of the monetary organs of the City. As a general rule, the writers on monetary matters in the leading London newspapers are well informed; they are men who are themselves able to read the signs of the times. But the epoch of which I speak was exceptional; certain organs of public opinion were deceived, and contributed unintentionally to propagate these monetary heresies.

Every one knows the rest. Symptoms of the great crisis first appeared in 1865. Many large houses in Mincing Lane suspended payment. Then followed some banks and some large India and China houses. The fever for bringing out public companies had somewhat abated—the first symptoms of the fatal epidemic that laid waste the shares of limited companies were subsiding—when a new class of men suddenly sprang up, and took the place erst occupied by the promoters. These new-comers very much resembled the cattle-plague doctors, who slaughtered the creatures they were called in to cure.

I can understand a lottery. You buy your ticket, and either lose your money or else win something. I understand the *roulette* of Baden-Baden; I can even comprehend the English turf, the Roman *tali*, and the Greek *astragalus*. In the hazard attached to these concerns you have a real, though not, perhaps, a very fair chance. But what chance has the outsider on the Stock Exchange? He can never buy stock, even for investment, and realize, at the same moment, the price paid. If he buys for speculation, from the moment he becomes the holder of stock he is every fortnight subjected to a fine, called *contango*—a heavy interest on the stock he holds. Persons who are in possession of good or bad intelligence with regard to the stock in question, or who are aware of a prepared rigging of the said stock, may, in spite of all these disadvantages, make a profit out of it. But are not such profits the result of a fraud committed by the few on the many? And are not such *tours de force* the daily practice of the members of that great institution? How can the unenlightened public fairly compete with the insiders in that mighty abode? Within these mystic precincts there sits a committee, absolute as any autocrat, and always ready to hurl some fresh edict against any outsiders who seem likely, as we say when playing at the game faro, "to blow up the bank." What chance have you, or any other outsider, against a body of men who can prohibit transactions in the shares of a public company, with which they had, perhaps, a few days before helped to saddle you! With the sharpest horse-dealers you have a better chance than with these Stock Exchange gentlemen. The former will not prevent you putting up to auction even the lame horse that they have imposed on you; the latter annihilate the shares for which you in good faith paid your money. The redress of this evil rests entirely with the Legislature.

CHAPTER XLIV.

THE ANGLO-GREEK STEAM NAVIGATION AND TRADING COMPANY.

ABOUT the middle of 1864 I began to retreat from the Stock Exchange. I was thinking seriously of establishing a new line of steamers, to succeed the Greek and Oriental, or the " Levant and Black Sea," as Messrs. Overend, Gurney, and Co. had re-baptized my line. Understanding as I did the movements of Messrs. Overend, Gurney, and Co., I felt that they could not remain much longer in the field. My friend M—— is to blame that this new concern did not appear in July, 1864.

After my separation from Messrs. Overend, Gurney, and Co., they embarked in a perilous and illegitimate grain trade, which, although it momentarily satisfied certain wants of theirs, and gave them an opportunity of making nice entries in their books, gave, as was inevitable in the long run, the *coup de grâce* to the line.

But, strangely enough, in the eyes of Messrs. Overend, Gurney, and Co., I was the great criminal, merely because I was the originator of the line. The second engineer on board a steamer once said to his *chef*, who uttered a malediction against the engine: " 'Tis not Stephenson's engine that's bad, Sir, but, the way you work it." This is a fair illustration of the Greek and Oriental Steam Navigation Company in the hands of Messrs. Overend, Gurney, and Co. That

T

line was firmly based; it rested on a large and lucrative trade.

The Greek and Oriental Steam Navigation Company expired about the middle of 1865, two years after I left it, and eight years after it had first sprung into existence. That company, in the meridian of its power, possessed twenty-four steamers; at its close it numbered seven. Of these five were sold by Messrs. Overend, Gurney, and Co. to a Liverpool merchant, at prices very little less than what I had originally paid —a proof that I had bought them cheap.

The offices of the company, in Fenchurch Street, were now abandoned to a host of clerks, whose most laborious occupation consisted in paring their nails, and who found recreation in discussing the merits of the various *chefs* under whom they had served in that office.

It was with no small share of grief that I witnessed the expiring throes of a line which, in the natural order of things, ought to have enjoyed a long and flourishing commercial existence. But, on the other hand, considering the turn events had taken, I felt glad that the Greek and Oriental Steam Navigation Company had, as it were, stolen out of life, and not come to an end amidst some public scandal, the odium of which its then powerful possessors might have contrived to throw upon me.

That my apprehensions were not ill founded was abundantly proved when I began to establish the Anglo-Greek Steam Navigation and Trading Company. This line of steamers was to have traded on commission and consignments alone; there was to be no speculation. And yet scarcely had I printed my prospectus when a secret but formidable opposition began to follow my paths. I had applied personally or through highly respectable men to several members of old-established firms. All found my project admirable, and several consented to become directors. But within a few days I was sure to receive a note declining the honour. The more I endeavoured to find directors of commercial weight, the more it become known that I was bringing out a steam company, the more terrible was the opposition I encountered. I received anonymous insulting letters, and I was told that my name was the sole cause of the company not floating.

I was more amused than discouraged by the host of enemies that I found springing up around. I felt that truth would eventually triumph. I knew that I had many lions in my path, but I also knew that the animals were lamed, and dared not face me. A man who seeks to justify himself by calumniating another, or who writes anonymous letters through revenge, is as little to be feared as believed.

Early in 1865 I again left the City, having failed to induce any of what are called "tip-top" men to join the directorship of the Anglo-Greek Steam Navigation Company. I know that Messrs. Overend, Gurney, and Co., and their favourites, had not spared me. I knew that they had not only hinted, but openly asserted, that trade in the line I proposed existed only in my imagination; that were it otherwise, with the facilities they had given me, I ought to have long before converted it into a monopoly. Were any of my acceptances brought to them to be discounted by a broker—and you know, reader, there was scarcely a City man whose acceptances did not pass through that great melting-house—they were refused in a damaging fashion. One partner, perhaps, passed the bill to another, with a significant smile or wink, or it was loudly refused, with, " No, no, thank you. Xenos—Xenos; we know that gentleman well." What could I do? I could not go round to every one who heard these injurious hints, and relate the lengthy story I have told here. Still less could I hope to meet a man of perspicacity sufficient to discern that the Cyclopean statue on which he gazed with awe was not indeed a divinity. The human race is essentially the same under all those social changes that we call " progress " and " civilization." In every age, in every clime, credulity, with the masses, takes the place of faith. The thoughtful Egyptian regarded the crocodile as a deity; the highly civilized Athenian looked upon the ugly owl as the harbinger of good news; the sagacious Theban drew favourable auspices from the convulsions of the dove struggling in the talons of an eagle. And no doubt posterity will discover in some of our current beliefs, follies as egregious as those that distinguished past generations. One man may, like Socrates, discern the truth amidst the darkness of this paganism; but would a modern Socrates have a greater chance of being listened to than had the ancient? I think not. Messrs. Overend, Gurney and Co. were the metal deities of the City. Their feet were indeed of clay, but those their adorers did not see. They looked only at the head. These terrible Gog and Magog ruled the money world. They could have crushed twenty like me. On the other part, in my letter of the 20th of May, 1863, to Mr. G. B. Carr, a copy of which I sent the same day to Messrs. Overend, Gurney, and Co., I told them that I hoped the Almighty, even if I should be left penniless, would prevent me from ever being forced to ask a favour from them, and that my accounts with them were closed for ever *in this world.* (See page 243.) I could not, therefore, all things considered, make up my mind to ask them to assist in establishing my new company.

I was one day speaking with Mr. Hollams about the Articles of Association of the Anglo-Greek Steam Navigation Company, when he said, in his usual quiet way:

"Do you know that you cannot bring out a steam company for that trade? You forget that you sold the Greek and Oriental to Overend, Gurney, and Co. I don't say you can't bring out a company—of course you can; but a short advertisement in the newspapers, to the effect that the Greek and Oriental has priority in the trade, would prevent the public from applying for shares; besides, the present owners of the Black Sea and Levant may stop the Anglo-Greek in Chancery."

"What's to be done, then? What do you advise?"

"Couldn't you see Carr, and try to come to some understanding with him?"

"Since March, 1863, when I broke off with the firm, I have never met any of them. Perhaps you can help me by seeing Carr, and sounding him."

"Very well. I have to see him about some other matters, so I shall take the opportunity to speak about yours."

After leaving the office of Thomas and Hollams, I began to reflect seriously on what Mr. Hollams had said. Knowing that the Greek and Oriental Steam Navigation Company was dying a natural death, I had, a few months previously, gone round to the principal Greek shippers, in company with another Greek gentleman, who was to join me in the board of the new company, limited, and had succeeded in inducing the shippers to sign a new letter expressing their intention to give me the preference of their shipments in my new line. This was a valuable concession. I had about the same time obtained, through the Greek Government, the special patronage of His Majesty the King of the Greeks for the Anglo-Greek Steam Navigation Company. In virtue of the royal patronage, the steamers of that company were to have many privileges, particularly at Pyræus, which was destined to be the great transit port for goods exported from England to the Levant, as well as for those imported to England. I also contracted with several ship-builders in the north of England for sixteen steamers, which were to form the above line, to be paid for partly in shares, and partly with the acceptances of the Anglo-Greek Company. The general scheme for working the Anglo-Greek Steam Navigation Company was as follows :—

The company was to be a trading one, and was to have several small steamers, of about 700 tons burden, for the transport of the mails and the collection of goods. These small steamers were to serve as feeders to the large ones, which were about 2000 tons burden. Now as to the working.

There are in the Levant ten principal towns whence the products of Turkey, Greece, Egypt, and Russia are exported, and where English merchandises are imported. These towns are—Odessa, Galatz, Constantinople, Salonica, Smyrna, Beyrout, Alexandria, Syra, Patras, and Corfu. Each of these towns is, so to speak, a transit port for the adjacent coasts, as well as for the interior, when there is not a direct communication with the great emporium. For example: Hundreds of small dealers are obliged to transport their goods from remote villages and small seaports on the coasts of Asia Minor and the neighbouring islands to Smyrna. The mode of conveyance is by small sailing craft, and the freight for these short distances is often equal to what is paid from Smyrna to London. The mode of operation is, that these goods are sold to the large exporters, or are consigned to them, to be reconsigned on account or joint account to England. Corfu is the commercial drain, as it were, for the coasts of Albania; Beyrout discharges those functions for the coasts of Syria; Salonica, for the coasts of Macedonia, Thrace, and Thessaly; and so on of the other large towns.

The plan on which I proposed to work was this:— Each of the small steamers was to leave Pyræus or Athens once a week, coast along one of the above-named shores, and receive the goods from the owners. Upon these goods our agents were to advance half or two-thirds of the value, and load them direct to London, consigned to the Anglo-Greek Steam Navigation Company, with one through freight. In London the goods would be sold on commission, or on the owner's account. The money-advances on such goods might be safely made in two ways; they might be made in cash through the captains, supercargo, and agents conjointly, or through the central banking agency at Athens, after the goods were shipped on board the steamers, and the bills of lading drawn to the order of the Anglo-Greek Company in London; or the exporters could have direct bills of exchange on London, or on the national bank at Athens.

It is quite plain that in this way we should not only facilitate the transport of goods to and from London, but we should also open fresh tracks of communication with the interior of these countries, and so enlarge the trade. In a country like Turkey, where the want of roads prevents a free internal communication, the punctual arrival of the steamer at these new seaports would supply a great deficiency; and these small steamers, returning to the Piræus on a fixed day, would bring, without competition, a full cargo for the large steamers bound to London. The exporters generally order the balance coming to them after the sale of the goods in England, to be converted into manufactured goods, colonials, or metals for exportation, consequently the outward cargo for the steamers would be secured without competition.

But, perhaps, I may be asked how this would affect the Greek firms of London, that gave me so liberal a support. I shall explain. The Greek commercial houses of London are divided into two classes —those that trade on their own account, and those that trade on commission. Those that trade on commission are few, and the great majority—those that trade on their own account—would give me their goods because my line would be a national one, and would tend greatly to increase Greek commerce. Even those who traded on commission would be compelled to ship with the Anglo-Greek Company, because it would be the only line having a direct communication with all the islands of the Archipelago, with the seaports of the coast and in-lying districts whence they received their orders. So that, under all aspects, this line, well worked and with sufficient capital, would secure a kind of monopoly in freights, independent of the 3 per cent. commission on all the goods consigned, bought, or sold for the consignee's account.

My next step was to form a board of directors for

the Anglo-Greek Steam Navigation Company. I resolved to select six London merchants, solvent men, whose character would be a security to the public when applying for shares. With such a board, and the patronage I was promised, I could feel no doubts as to success. But when I went into the City and made offers of directorship to the men upon whom I had reckoned, I soon found how deep was the prejudice entertained against me on account of my transactions with Messrs. Overend, Gurney, and Co. My position was a trying one. I knew I was slandered, I knew that groundless prejudices were fostered to my disadvantage, and I was compulsorily tongue-tied.

Frustrated in my search for directors, I began to revolve in my mind the advice given by my friend, Mr. Hollams. I must here avow that in laying down the plan for the Anglo-Greek, I had wholly overlooked—in fact, had quite forgotten—the impediments which the purchasers of the Greek and Oriental could throw in my way. In those days Messrs. Overend, Gurney, and Co. wielded a power in the commercial world which no individual could resist. Their frown alone was sufficient to blight. But, independent of their monetary and moral weight—in those days it was moral—they had, in the existing state of things, a legitimate power to check me.

The more I reflected, the more I felt convinced that the advice of Mr. Hollams was the best I could follow. I saw that the great capitalists had turned the Millwall Works into a public company with a capital of several millions sterling, and I had no doubt of their having done equally well with a great deal of the rubbish they held. I saw that the Crédit

Foncier et Mobilier had brought out a large steam navigation company, with fleets once possessed by Messrs. Overend, Gurney, and Co., and supposing that by these transactions the great firm of Lombard Street must have realized pretty large sums, I began to believe they had tided over their difficulties, and would not be sorry, without appearing openly in the affair themselves, to help Mr. G. B. Carr and me to establish this new line upon a solid basis.

Whilst in this frame of mind I one day accidentally met Mr. A. Carnegie, who had just come from Galatz to commence business as a merchant in London. In the course of conversation I mentioned that I was about bringing out a new line to succeed the rapidly dying Greek and Oriental. Having dilated a little upon my projects, and the difficulties I had encountered, Carnegie offered to induce Mr. Carr to procure me the moral support of Messrs. Overend, Gurney, and Co., if, in return, I would promise his house at Galatz the shipment of grain for the new company.

Two or three days had passed since Mr. Hollams had promised to see Mr. Carr, and the interview had not yet taken place. I was getting impatient, and as the new company was to work on commission and not on speculation, I thought I could not do better than accept Carnegie's offer. I did so, and he immediately set to work.

The negotiations between Mr. Carr and me were very protracted. He declined to become a director in my new company. Messrs. Overend, Gurney, and Co. refused to give me any moral support, to take shares, or find me directors—in fact, to sign the contract drawn by A. Carnegie, marked No. 29 in the Appendix. All this parleying consumed time which was all important to me, and, as I was anxious to remove all impediments that obstructed the coming out of the Anglo-Greek, I signed, on the 25th of April, 1865, a contract for the purchase of the goodwill of the Greek and Oriental Steam Navigation Company for £3500, Mr. G. B. Carr representing Messrs. Overend, Gurney, and Co. in the transaction.* I had also to pay the salaries of the clerks from this date up to the day that the company was brought out.

The failure of the Anglo-Greek Steam Navigation and Trading Company, Limited, was due to three causes. Firstly, because I did not succeed in my attempt to induce substantial City men to join the board of directors. Secondly, because I brought my company out at the eleventh hour, when the public craving for limited companies was already satiated; consequently the shares were not largely subscribed for, and a sufficiency of capital could not be raised: And thirdly, the company did not succeed because it was LIMITED. Whoever reads my letter to Mr. Henry Edmund Gurney, written in 1860 (see page 77), will perceive that I foresaw the evils inevitably attendant on the limited system. That system was developed, so to speak, under peculiar conditions of the commercial atmosphere; and the fact that, of the hundreds of limited companies that sprang into existence, so few reached maturity, may be adduced as a proof of the truth of my predictions.

---

* This contract, being one of the concessions which I sold to the Anglo-Greek Steam Navigation and Trading Company, Limited, I delivered to the company's solicitor, Mr. Sharp. Owing to his lamentable death, I am unable to produce it, but the receipts, marked Nos. 42, 43, and 44 in the Appendix, show the validity of the transaction.

That the Anglo-Greek Steam Navigation and Trading Company was based on sound principles, and could have been worked to yield great profits, I have had substantial proofs. Were it otherwise, would Mr. Francis Burnand, an eminent stockbroker, have offered me, on the part of certain great capitalists, £25,000 for my concessions? I should say certainly not. My reason for not accepting that offer was, that by the conditions of the bargain I was to withdraw altogether from the company, and be neither manager nor director. But as the patronage of His Majesty King George of Greece had been granted to me personally, I deemed it a point of honour to remain identified with the company. I now only mention this offer of £25,000 to prove that if the Anglo-Greek were not a promising enterprise, shrewd men of business would not have esteemed it worth that price. And there is the opinion of the Master of the Rolls, when, in 1866, the Anglo-Greek was brought judicially under his notice. His lordship on that occasion examined and admitted the validity of my concessions and the well-doing of the company, and therefore dismissed the petition of Messrs. Grant and Gask.

And yet, of all connected with the Anglo-Greek, I have lost most. I sacrificed my house—Petersham Lodge—and its valuable contents, to promote the company. I was the largest shareholder, having nearly 3000 shares, which I kept to the last—a proof of my faith in the vitality of the Anglo-Greek. My conduct was investigated before the Master of the Rolls, and I can confidently refer the reader to the official liquidator, Mr. Henry Chatteris, who with the solicitors of his firm, Messrs. Davidson, Carr, and Banister, examined the accounts of the company, and read my protest on the minute-books. I owe it to the integrity and kindness of these gentlemen, and to the high sense of justice of the chief clerk, Mr. Church, as well as of the learned Master of the Rolls, that I escaped the Bankruptcy Court. It would have been dealing me the *coup de grâce* to have sent me to meet there, after so long an interval in our acquaintance, my old friend, Mr. Edward Watkin Edwards.

CHAPTER XLV.

THE CONSULATE.

THE Fata Morgana seemed to preside over my destinies during the year of grace 1865. Everything looked well, looked brilliant; everything resolved itself into mist and vapour. The Greek Government, on the 5th of June of that year, appointed me Greek Consul-General in London. The British Government, without assigning any reason, refused my *exequatur*. I never have been able to learn the cause of this proceeding. During twelve months the post of Greek Consul in London was virtually vacant, and my Government used every effort to persuade the British to accept me as consul; but in vain. The discourtesy to the Greek Government was persisted in, and the persecution of an individual was not considered unworthy of a mighty Government. However, a better state of things is now being inaugurated for England. The late revelations with regard to the long-boasted commercial honesty of her mighty merchants will not, it is to be hoped, prove unfruitful. Englishmen may there learn a modest diffidence when, in future, they indulge in eulogiums on what they are pleased to call their "national virtues." They may also learn to treat with forbearance, perhaps ultimately with justice, foreigners who have paid dearly for trusting too largely to English commercial probity.

Let no one accuse me of speaking with bitterness.

A man deeply wounded cannot be expected to break forth into mirthful songs when he is cramped with pain. The refusal of the British Government to grant my *exequatur* was not alone mortifying to my feelings; as a simple fact, it became in the hands of my enemies a new source of misrepresentation. A rumour was circulated by malicious people that the Greek Government asked me to resign because of the Chancery suits connected with Messrs Grant aad Gask and the Anglo-Greek Steam Navigation Company; such was not the truth. I was appointed consul before the Anglo-Greek "floated," and long before it appeared in Chancery the determination to refuse my *exequatur* had been arrived at.

A strong refutation of all these calumnies will be found in the fact that for nearly a year my Government persevered in trying to induce the British Government to yield in my favour. These efforts proving useless, another gentleman was appointed consul. The following translation of a letter of the then Greek Minister for Foreign Affairs—Mr. Spiro Valaoritis—to my father, will confirm what I say:—

[*Translation.*]

Corfu, $\frac{16}{28}$ *January*, 1867.

MY DEAREST MR. XENOS,

I received your dear letter of the 7th inst.; also the copy of your son's letter. I have perused both attentively, and am much grieved to learn from your son's statements that he has suffered greatly, and been unjustly persecuted by men from whom he had a right to expect gratitude.

I have acted towards him frankly and unreservedly. I advised him, through you, to resign, because that my Government was not able to obtain for him, from the English Government, his *exequatur*. The English, like every other Government, has the right to refuse or grant such. May an opportunity occur to enable your son to clear himself from the slanders that have been uttered against him.'

Wishing you happiness,

I am,

Your friend,

(Signed)   Spiro Valaoritis.

The slanders above alluded to were the evil rumours circulated to my disadvantage by certain persons living in England—persons, too, from whom, if they were as patriotic as they pretend, I ought to have received very different treatment. These people succeeded so far in their malice as to pour their evil-speaking into the ears of the noblemen and honourable gentlemen who then ruled in the British Foreign Office. I cannot refute what my enemies said against me, as I have never learned the character of their charges or inuendoes; but this I know—they succeeded in attaining their object.

Now, at the end of three years, I can speak calmly on the subject, because in that short space so many bitternesses have been crowded into my existence as to push, so to speak, the circumstances connected with the refusal of my *exequatur* far back in my memory. In this spirit, I cannot help expressing my surprise that the very nationality of my enemies did not make them suspected by men experienced in the ways of the world, and accustomed to observe the thousand meannesses that disfigure even comparatively good characters. A distinguished modern historian says something to the effect that a successful man is sure to find his bitterest enemies amongst his own townsmen. A number of foreigners living in a large metropolis form, as it were, a micropolis, within which live and flourish myriads of petty village bickerings and jealousies, the expression of which should never be allowed to influence the judgment of the great governing powers.

As for myself, I do not think I deserve badly of the British Government. During twenty years I wrote and spoke in favour of British preponderance in the East; I laboured unceasingly to introduce British institutions into those regions; I raised my voice against the wild prejudices engendered by Panslavic teachings; in fact, if I have sinned politically, it has been through a desire to make Greece somewhat English. I was the first who proposed that the crown of Greece should be offered to Prince Alfred of England. I formed a Philhellenic Committee in London, the object of which was to improve and maintain the friendly relations existing between England and Greece. I also endeavoured hard in 1863 to effect the settlement of the vexed question of the Greek bonds of 1824-25. And with regard to these bonds, I tried to make Greece adopt a policy becoming the grandeur of the future which I hope is before her, and which would have secured to her the respect and confidence of her English creditors.

If I have made enemies and excited slanderers by my endeavours to strengthen and cement the amicable relations existing between England and Greece, I must be content to bear the consequences of a line of action followed in obedience to conscientious convictions. But, whilst I can understand and know how to despise the short-sighted envy and ignorant malice of men who murmur at the elevation of one of their own compatriots, because he has happened to wound some of their political prejudices, I certainly cannot comprehend how British ministers could allow themselves to be made instruments to gratify a paltry hate. It is only a truism to say that *ex parte* statements should be received with caution, or that justice demands the patient hearing of both parties in a suit. I advised the honourable settlement of the question of the Greek bonds of 1824-25, and by that, as well as by other advice of a similar character, I offended interested debtors, without, as it would appear, gaining the goodwill of the creditors I tried to serve. But the day will inevitably arrive when every enlightened Greek, as well as every honest Englishman, will acknowledge that, had the question of the Greek bonds been settled in 1863, the aspect of affairs in the East would be pleasanter than it is at the present moment.

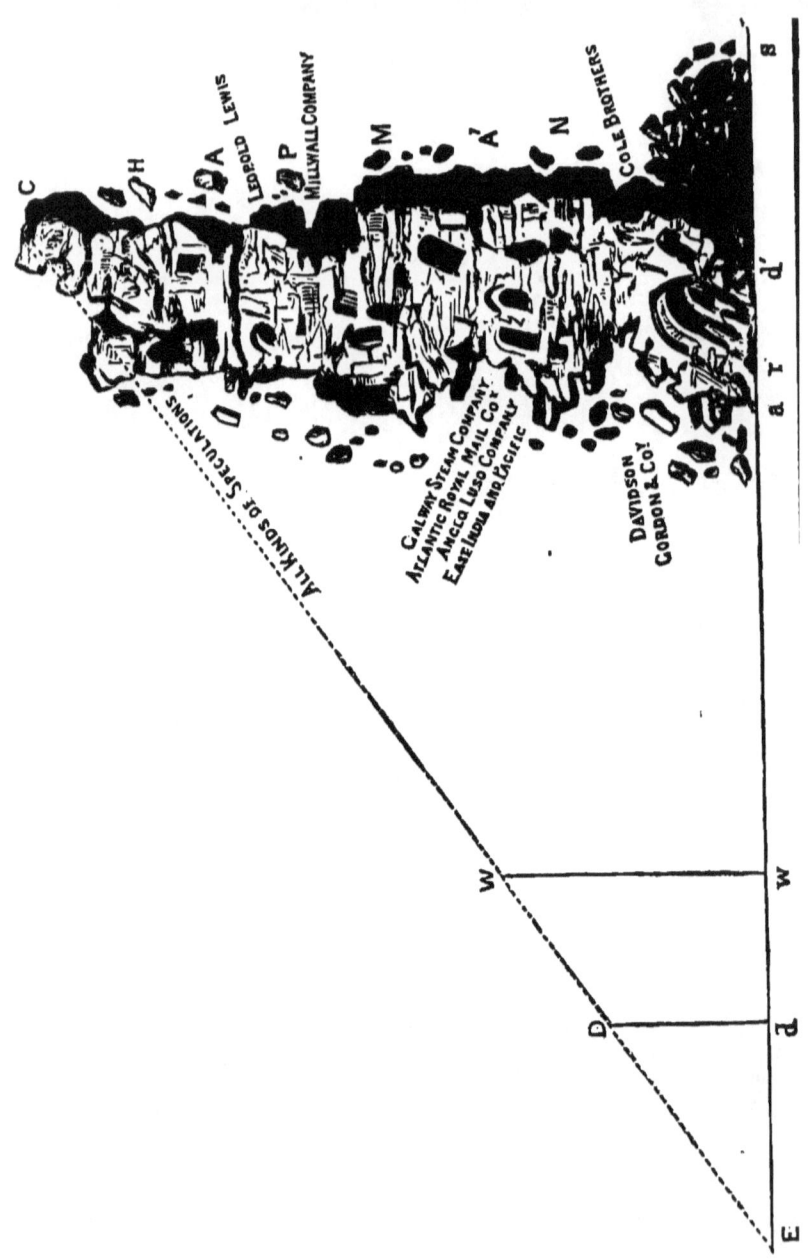

CHAPTER XLVI.

DEPREDATIONS.

I SOLD the Greek and Oriental Steam Navigation Company on March 31st, 1863, for **£2500**; I bought it back on April 25th, 1865, for **£3723**. (*Vide* Appendix, Nos. 42, 43, and 44.)

These two sums, put in an algebraic form, give the following result:—

$$£2500 - £3723 = - £1223.$$

This little sum of — **£1223**, paid out of my own pocket, comes under the head of what Mr. David Ward Chapman calls, in his letter (Appendix, No. 51A), my DEPREDATIONS. However, to ascertain the full extent of the depredations operated in this commercial Tower of Babel, we have to solve the problem represented by the annexed diagram.

We want to ascertain the perpendicular height of C.H.A.P.M.A'.N.s. This altitude not being within the reach of manual measurement, we proceed by reckoning to attain our object. Having measured the base line E.d.w.a.r.d'.s., and marked off the two triangles, E.D.d. and E.W.w., and drawn a line of connexion between C. and E., we find the following results, in accordance with the principle of the similarity of triangles and the relative proportions of parallel lines:—

Ed : Dd : : E.d.w.a.r.d'.s : C.H.A.P.M.A'.N.s.

That is, by multiplying the second and third terms, and dividing by the first, we have the full altitude of C.H.A.P.M.A'.N.s.—thus:

$$\mathrm{C.H.A.P.M.A'.N.s} = \frac{\mathrm{E.d.w.a.r.d'.s.} \times \mathrm{Dd}}{\mathrm{Ed}}$$

That is to say, the altitude of C.H.A.P.M.A'.N.s. depends on the length of E.d.w.a.r.d'.s.

CHAPTER XLVII.

THE RATS AND THE PIG.

DAVID BARCLAY CHAPMAN, father of David Ward, is afflicted with a feeble memory. He forgets Joseph Windle Cole, who, emerging from the counting-house of Forbes, Forbes, and Co., in which he was a shipping clerk, traversed Eastern India in various capacities, and with a rapidity that would have done credit to one of the genii of the "Arabian Nights," in 1847, presented himself in Basinghall Street, loaded with the *heroic* liability of £153,000, and who in the following year—1848—established himself in Birchin Lane, where he puzzled his mighty neighbours, the denizens of the "Corner House," with fictitious warrants of the Hagen's Sufferance Wharf, and his astounding cash account of two years, amounting to £4,300,000, together with his new estate in Basinghall Court, against which Messrs. Overend, Gurney, and Co. had a claim of £120,000.

David Barclay Chapman also forgets Charles Maltly, Esq., who was a Custom-House clerk in the firm of Messrs. Forbes, Forbes, and Co., and who, like Joseph Windle Cole, left that employment; but he moved at a slower pace, and journeyed through more obscure ways than did his fellow-clerk, until he ultimately became the illustrious Maltly and Co., wharfingers, of the Hagen's Sufferance Wharf—a locality that attained an equal celebrity in connexion with fictitious warrants, that entailed no small amount of debt on the great capitalists.

David Barclay Chapman forgets Mr. Sargant, another clerk of Messrs. Forbes, Forbes, and Co., who, after leaving their house, established himself in a business-like manner, in Mincing Lane, as Sargant, Gordon, and Co. David Barclay Chapman forgets how this establishment came to grief, or, in other words, became insolvent in the same year, 1847, with liabilities amounting to £62,254 1s. 5d., and how from the ashes of this firm sprang up another, Davidson, Gordon, and Co., a concern redolent with the odour of fictitious warrants, secured to Messrs. Overend, Gurney, and Co., with a profitable distillery in West Ham, in the county of Essex.

David Barclay Chapman forgets the two fugitives, George Sedgwick and Charles Gray, and their adventures at Neufchâtel with Madame Formachon, when Mr. James Beard, of Beard Brothers, Manchester, was tracking them with detectives. He forgets, too, the display of daggers and revolvers made by these gentlemen at Naples.

David Barclay Chapman forgets the disclosures made by Mr. William Bois, his confidential clerk, on the 31st of May, 1856, in his affidavit, where he says that on the 13th of October, 1853—that is, nearly four years before the bankruptcy of J. Windle Cole—a conversation occurred between the latter and Mr. David Barclay Chapman relative to certain dock warrants held by Messrs. Overend, Gurney, and Co., and which warrants purported to represent spelter, tin, copper, Swedish iron, lead, tin plates, and cochineal, lying at Hagen's Sufferance Wharf, to the amount of £250,000, and against which warrants Messr.

Overend, Gurney, and Co. had advanced to Cole Brothers upwards of £195,655; and that in the aforesaid conversation Joseph Windle Cole admitted to David Barclay Chapman that the aforementioned warrants were valueless, the goods which they purported to represent not being at the wharf.

David Barclay Chapman forgets the strong remarks made by the Recorder, the Right Hon. Stuart Wortley, when he said—" I shall not anticipate anything, lest I might be the means of doing injustice; but I will say that I believe it to be unfortunate that those who had a knowledge of the position of those parties, in the earlier stage of their proceedings, did not take means to stay them. It may have been from motives of kindness, but I think they were mistaken motives, and that they were the means of inflicting injury upon other parties." *

David Barclay Chapman forgets the lamentable death of a benevolent and honourable man, Mr. Samuel Gurney, his senior partner, a man whom the exposure of the above-mentioned transactions brought to his grave. He forgets that these transactions, during the time he had the management, and their consequences, made the first breach in the great tower of Overend, Gurney, and Co., sapped its foundations, and precipitated its fall.

Forgetting all these antecedents, David Barclay Chapman thinks proper now to step before the public and express his " deep affliction," for what has lately come to light. (See Appendix, No. 51.) He publishes a letter addressed to him by his dear David Ward, which he " believes to be perfectly true," and

---

* Before Mr. Commissioner Fonblanque, July 2, 1857.

adds that he thinks it his duty to speak upon a point in which the "highest moral principle is involved."

In what strange times we live! Who that has read the life of Ali-Pasha of Janina would expect to find off-shoots of his school in London. We can only repeat the common Greek saying: Παντοῦ τὰ πάντα.

What an admirable production is the letter addressed to David Barclay Chapman by his son! (See Appendix, No. 51A.) In that epistle Mr. D. W. Chapman summons the whole world into the presence of his Maker, to show where he has sacrificed the interests of Overend, Gurney, and Co. The reader of "DEPREDATIONS" will be able to form an opinion for himself as to who sacrificed the interests of Overend, Gurney, and Co.

David W. Chapman thinks it was necessary to keep out of the Bankruptcy Court the dreadful accounts of Lever, Mare, and Xenos.

If these accounts were really subjects for the Bankruptcy Court, why did the Overends, by the advice of Edward W. Edwards, and under the management of D. W. Chapman, buy up these concerns, and carry them on for several years afterwards, appointing their own favourites as managers? Was not Mr. J. Lyster O'Beirne an acquaintance and private friend of D. W. Chapman? Was it not D. W. Chapman who introduced him to Mr. John Orrell Lever, and afterwards appointed him to manage Mare's business? Was Mare's business improved and its debt reduced under this new *régime*?* Was not Mr. Edward Watkin

---

* I do not allude to any moneys which, in 1864, the Millwall Iron-Works Company may have received from the public when the concern was converted into a limited company with 10,000 shares and £2,000,000 capital. Mr. J. Lyster O'Beirne, in his letter to

Edwards an old and special friend of Chapman, when he was appointed, with a gift of 100 shares, general disposer of the fate of the Atlantic Royal Mail Steam Navigation Company? Was it not under his management that the debt of that company rose from £200,000 to £850,000?* Why was the Greek and Oriental, after I sold it to the " Corner House," transformed into the Black Sea and Levant Steam Navigation Company, and allowed to trade two years longer in grain speculations?

I challenge the whole world, not into the presence of my Maker, but into the presence of these facts, and ask whether any one can there show that D. W. Chapman did not sacrifice the interests of Overend, Gurney, and Co., if not directly in his own person, at least through the instrumentality of his favourites.

When a large house like that of Overend, Gurney, and Co. takes up certain working concerns, and carries them on for a succession of years, there are only two modes of explaining such procedure. Either the concerns were sound at the time of being purchased by the capitalists, and were brought to ruin by their mismanagement; or else, the purchasers were themselves insolvent, and sought to make a scapegoat of their

---

he *Times* (see Appendix, No. 52), says that in the early part of 1861 he was appointed by Messrs. Overend, Gurney, and Co. superintendent of the Millwall Works, with a salary of £1500 per annum. Let Mr. Mare say what his debt was at that date, and if it was anything approaching the sum that figures in the present estate of Messrs. Overend, Gurney, and Co., and if the Millwall Works were a concern for the Bankruptcy Court, why they carried them on for three years?

* Mr. John Orrell Lever, in his Letter to the *Times*, states that, when he left the Galway Company in the hands of Mr. Edwards and Mr. D. W. Chapman, this concern was perfectly solvent.

late purchases by throwing on them the odium of ancient losses. This would be equivalent to committing DEPREDATIONS on the public.

But D. W. Chapman, in his published letter, makes an admission equivalent to saying that Overend, Gurney, and Co. were insolvent, when he says that it was in the power of Edward Watkin Edwards to make them put up their shutters within four-and-twenty hours. Can a banker or merchant be so deeply in the power of his servant or *employé*, unless he is insolvent—or something worse?

D. W. Chapman says he had much difficulty in dissuading Edwards in 1861 from asking for a partnership in the firm of Overend, Gurney, and Co. No man of business will believe that assertion. Mr. Edwards would not be so great a simpleton as to embark on board a vessel that he knew was drifting into a whirlpool.

D. W. Chapman hopes the explanations which he gives in his letter to his father may be satisfactory; and he adds that, whether they prove so or not, his conscience accuses *him keenly of many most culpable follies in his private capacity*, but of disloyalty to the house, never.

Eminent writers on spiritual questions tell us there are many kinds of conscience, and warn us especially against a FALSE CONSCIENCE.

D. W. Chapman talks of me as one of his bitterest enemies, because, as he says, it was he who tried to stop my depredations. I have told Mr. D. W. Chapman, both verbally and by letter (see pages 242, 243), that I am ready to defend my character to the last drop of my blood. Why did not Mr. D. W. Chapman, on either of these occasions, use the word "depredations"? Why did he not take legal steps to punish me? However, whilst the gentleman remains ensconced at Tours, I can only say, Οὔ σὺ λοιδορεῖς με ἀλλ' ὁ τύπος.

It must be regretted that Mr. D. W. Chapman has no adviser that could induce him to come to England at the present moment, when his character here is at stake, and where his presence in the witness-box is so much desired, in order to prove who is the real author of the DEPREDATIONS. By coming, he might aid his ancient partners to clear some doubtful points, and in this way he might make some atonement for "culpable follies" that cost a great deal of money. No matter what his commercial difficulties are, an order from the Bankruptcy Court would afford him protection against his creditors. Between two such evils, a man of honour would not hesitate which to choose.

I am not, and never have been, a bitter enemy of Mr. D. W. Chapman. I now repeat what I have often said—and I speak solely for the vindication of my character—I never flattered him; I never flattered his partners. I conducted all my transactions with Messrs. Overend, Gurney, and Co. in a strictly formal, business-like manner. Since 1862, I have not met any of the Gurneys, and have not spoken with any of their relatives, nor with R. Birkbeck; so that the opinions I now put forth are wholly without bias. I religiously kept my word not to trouble them. I faithfully observed my contract (see the fifth paragraph of No. 31 in the Appendix) not to print or publish anything that could injure—viz., an account of our mutual transactions; and by these scrupulous observances I seriously injured my interests, allowing myself to be crushed by the weight of their name, and the falsehoods propagated by their unprincipled favourites and satellites.

Mr. D. W. Chapman cannot say I came to this country an adventurer. I had, and have still, in Greece, a position. The name I bear is identified with the history of my country, by the patriotic sacrifices made by my family, and by their steady exertions in the cause of Hellenic independence. For myself, my fellow-countrymen in England have always treated me with consideration and esteem. I have always been a member of their patriotic committees, and have received many flattering testimonials from the Panhellinium, as well as from different Governments of Greece. It was a dark page of my destiny that opened to me an acquaintance with the Overends. On that fatal day I was arrested in the midst of a prosperous career, and ultimately cruelly sacrificed by men who brought ruin on thousands; and yet, spite of the anger and vexation consequent on my own treatment by the firm, I candidly declare my belief that the Gurneys and Birkbeck are greater victims than I. They had latterly become tools in the hands of unscrupulous and daring men, to whom, in an inauspicious hour, they had granted their confidence—men who, at the eleventh hour, knew how to evade the impending catastrophe.

But to return to the *conscientious* letter of D. W. Chapman. He is astonished how the new firm could manage to lose, during its short term of duration, the sum of £1,500,000. I shall endeavour to enlighten his mental obscurity by narrating a popular Eastern fable.

There is an island in the Greek Archipelago called Tinos, and on this island there once lived a pig so enormously fat that the rats came and burrowed in its back, and feasted luxuriously, without the poor pig being at all conscious of the DEPREDATIONS that were being committed on its body. At length the rapacious vermin, having gnawed, and nibbled, and eaten a way through the exterior coatings of fat, reached the sensitive flesh, and then the pig became conscious of the presence of the DEPREDATORS. The poor animal writhed in agony and grunted loudly, and the rats knew instinctively it was time to be off. So away they scampered. Some returned to their holes, others took to the water and actually reached the Continent, where they were beyond the range of the trap. Meanwhile the poor pig, though freed from its enemies, was past recovery, because the wounds inflicted by the crafty rats had penetrated to the vital organs, and, spite of a temporary but only apparent improvement, it very soon died.

Eastern fables sometimes serve to illustrate Western practices.

CHAPTER XLVIII.

CONCLUSION.

I HAVE now finished the task which a stern necessity imposed upon me. I did not seek publicity; it was thrust upon me. My name was incidentally introduced into courts of justice, casting upon me a darkly-shaded notoriety. My concerns were bandied about in newspaper columns as though they and I had only a mythical existence. If it is legitimate in those who win to laugh, I must say it is not at all pleasant for those who lose to be laughed at. The Greek and Oriental Steam Navigation Company was no myth, and to prove that publicly was a duty that I owed myself. The discharge of the duty thus thrown upon me has not been wholly free from difficulties. I had to walk along a narrow and sinuous path of self-defence, careful to avoid, on the one side, overhanging rocks, bristling with the pointed cannon of the law; equally anxious to shun, on the other side, the thorns and brambles that might wound the self-love of others, even when those others were my enemies. Should any doubt my tenderness towards my foes, I can tell them that, if I have said much, I have left much more unsaid. What I now publish has long been written—since 1863—and formed portion of a work which was intended for publication when the heat of personal prejudice and the warmth of personal resentment should have died out. Recent events have, so to speak, dragged me into print. I now stand before the tribunal of public opinion, not either as a plaintiff or defendant, but as an impartial witness of the truth. I have kept before my eyes the delicate position of the defendants in the impending trial, and though I have nothing to say that could prejudice their cause, I have sedulously foreborne to advocate their case before an injured public. My first and great aim has been to vindicate my own character. Having, as I hope achieved that object, I may be permitted to repeat, as I conscientiously can, that the Messrs. Gurney and Birkbeck have been victimized. It is much to be regretted, both for public and private reasons, that men of their high social standing and moral life should find themselves under the cloud of a suspended judicial examination. This simple expression of opinion will not be deemed flattery in one who has suffered much from the firm of Overend, Gurney, and Co.

The following are the steamers that formed this line from 1857 to 1865:—

The Greek and Oriental Steam Navigation Company.

LARGE STEAMERS.

	BURDEN. TONS.	H.P. COMBD.
General Williams	2000	250
Admiral Miaoulis	2000	700
Palikari	2200	600
Mavrocordatos	2200	600
Scotia	2200	500
Asia	2200	500
Admiral Kanaris	1500	400
Modern Greece	1200	400
Powerful	1200	450
Marco Bozzaris	1000	400
Petro Beys	1000	400
Lord Byron	1000	400
Leonidas	1000	400

DANUBIAN STEAMERS.

	BURDEN. TONS.	H.P. COMBD.
Tzamados	500	120
Botassis	452	100
Bobolina	300	80
Zaïmis	500	120
Colocotronis	500	120
George Olympius	500	120
Londos	500	120
Rigas-Ferreos	500	120

AMSTERDAM AND ROTTERDAM STEAMERS.

	BURDEN. TONS.	H.P. COMBD.
Smyrna	600	200
Patras	450	150
Colletis	500	120

CHARTERED STEAMERS.

	BURDEN. TONS.	H.P. COMBD.
Britannia	1000	300
Milo	1300	300
Hercules	1400	300

THE CITY OF LONDON

IN THE

PRESENT DAY.

The City of London is the great university where the human heart and the human countenance may be studied in their most striking varieties. It is, in fact, a vast school of moral anatomy, where human nature is laid bare, by the workings of its own passions, to the eye of the keen scanner. No matter how many the winters that had passed over your head, no matter what your experiences might have been in other parts of the world, you would be obliged, upon a short acquaintance with this great Pananthropopolis, to confess that all you knew was as nothing compared with what you might learn here.

The English metropolis is, indeed, a great school of moral anatomy, but the student has no occasion for a spatula. He needs only a keen eye to observe how the subjects reveal their structural capacity. And what struggles may not be discovered! What workings of nerve and muscle, of heart and brain! And all these nerves and muscles, all these hearts and brains, obedient to the idiosyncrasies of each individual, are still striving after one object. And what is that? MONEY, MONEY, MONEY. Money is the great climax that tops the ambition of every man who works in the City of London. And no wonder. What other possession is so honoured in our great Pananthropopolis?

Nay, make no remonstrance. I know, reader, what you are about to say. Talent is honoured. To be sure it is *if it be marketable.* But where does lean virtue make a more sorry appearance, beside fat and richly-robed wealth, than in the City of London? Oh, my friends, my friends! of a verity, money is the god of modern civilization, and modern civilization is best represented by prosperous commerce; and where does commerce find its fullest expression, its most ample exponent? In the City of London, I believe. Now, do not suppose that I am about to preach a long sermon, and find fault with all this. Not at all. I take things as they are—I merely assert that they are so; and, with your permission, reader, will observe that it is no wonder that a stranger, on his first arrival in London, is easily tempted to enter the vast arena where the commercial gladiators are engaged in the "keen encounter" of their wits, and try what he can do. This is provided he has a little wit.

I have spoken of the study presented by this fierce struggle for money, but I must further add, that in order to become a discriminating observer, a man must begin by being a combatant. In the pugilistic ring a few hard blows have the effect of shutting up a man's eyes; but in the commercial conflict it is quite the contrary—a few hard knocks improve the vision wonderfully.

Having been a combatant, and having received

some hard knocks, I will give, in general terms, some of the fruits of my observation.

Mr. Andrew W—— inherited from his father a small but respectable and profitable business. The father had traded after the old style. His mode was safe, but not brilliant. He preferred reasonable but secure profits, to grand possibilities attended by great risks. In short, he was a man of the old school. I need not remark that the present mode of doing business in the City of London is very different to that which obtained forty years, or even twenty years ago. Many a stately commercial fabric of the present day is based solely upon paper. We belong to a "fast" generation. I suppose the rapidity of our trade-pace over that of our fathers may be estimated by the difference that exists between the specific gravity of gold and paper. A friend of mine observed, some time ago, that a sound paper currency was, in the monetary world, a representative monarchy; indiscriminate bill-drawing, democracy with universal suffrage; and solid gold, a well-grounded and concentrated despotism.

But to return to my subject. The son, who, during his father's life, had been obliged to conform to an old-fashioned *régime*, resolved, upon coming into possession, to do a dashing trade. His intention was soon known, and a commercial man recommended him Mr. H——, as a head clerk, who knew how to do business according to the spirit of the times. Mr. H—— soon proved himself worthy of the recommendations he had received. He was soon made a partner in the firm, and secured in a handsome share of the profits. Mr. Andrew W—— gloried in the splendid prospects he saw opening before him. He drew bills and signed cheques for large sums, and marvelled at the dexterity of his partner in finding the means to meet all these liabilities. But it was all right. The account at his banker's was of five times the magnitude it had ever been during his father's life. His actual capital was five times, and his representative capital fifty times greater than when he became head of his house. His name was mixed up with large transactions, and he saw his firm shining like a comet in the *stereoma* of the City. However, he soon found that bills are not really what they represent, or rather, seem to represent; and, with all this fine show, very little money found its way into his pocket. Complications arose, confusion ensued, and poor Mr. Andrew W—— became a bankrupt. A few months previous his partnership with Mr. H—— was dissolved, and, strange to say, from the ashes of this bankruptcy his partner rose like a phœnix. *H*—— had made his MONEY. It is better not to ask how.

Let us now pass to the second act of this drama. The original proprietor became a clerk in the very office where he was once master. I have seen him gliding humbly and rapidly along the streets of the City, looking like a ghost of a past generation, whilst his *ci-devant* partner drove in his brougham to his West-end house. But you will say public opinion condemned him. Nothing of the kind. He associated familiarly with the very gentlemen who, when the firm became bankrupt, accused him of twenty times the amount of crime he had actually committed. And what is still more strange, these very people coupled his name with the epithet "clever."

However, in this particular case, I saw the reverse

of the medal. I saw the victim, Mr. Andrew W———, in the course of time, and by the assistance of friends, again lift his head. And I saw H———, the successful rogue, forget his craft, and gamble away his money on the Stock Exchange and in a life of dissipation. And when he in his turn became poor, the world, with wonderful practical sagacity, discovered that he was a rogue. He is now knocking about the City in rags, smelling strongly of liquor, as a *general broker*, happy if he can earn a shilling a day.

Mr. D———, a penniless, swaggering, pompous talker, took a furnished office, in 1855, in Fenchurch Street, and put his name on the door, with the addition of "Brothers." He had five apprentices, who acted as clerks, and who paid instead of receiving a salary. He had so far presented a front to the world. He had previously no hair on his face, so he grew a large beard and moustache. His manners are not bad; his assurance is boundless. He stepped one morning into the counting-house of my friend Mr. R———, a rich Greek merchant. He proposed a tremendous transaction with a certain great English firm, Messrs. D———, S———, and Co., being, as he says, a relative of John D———, one of the partners. Fortunately for him he has the same name as one of the partners, but not even the slightest connexion. I was present, but I could not recognize him or remember his name, on account of his beard.

Mr. D———, after sounding the inclinations of my friend, and ascertaining his terms, introduced himself with greater confidence to the English firm which he boasted to us—that is, that of his relative—and made abundant use of the name of Mr. R———, and several other Greeks. He commenced negotiations, and when his wonderful business was concluded, he must have made a commission of £1500.

This mode of proceeding presupposes shrewdness, a knowledge of the mercantile world, and unblushing effrontery, which is not original. After his first success, Mr. D—— had two respectable firms to push him on. He soon established a certain amount of credit for himself, and his office assumed an appearance of business. Two years after, Mr. R—— and Mr. D—— found out, for the first time, in conversation, that that special transaction was the first turn given to D——, the broker. It now depends on himself to keep his respectable position. Mr. D—— is now a very rich man, and a large merchant. He drives regularly, with his wife, in the season, in a large barouche. The English firm D——, S——, and Co., became bankrupts, but Mr. D——, the broker of this firm, is our worthy merchant.

So much for commercial stability. Shall I make a picture of a wolf in sheep's clothing? The schemer, in this case, Mr. Ch——, is deep-laid, and requires time for the working out. Our wolf fixed his eye on a grave and well-to-do Mediterranean merchant. He tracks him to his home, takes up his residence in the neighbourhood, comes to town by the same omnibus or train, and frequents the same church on Sundays. This constant propinquity ends in a kind of neighbourly acquaintance, In the journey to and from town, politics are talked; on Sundays, the sermon or the religious proclivities of the preacher form the subject of conversation. The wolf always agrees with his friend. This kind of out-door acquaintance continues for some months. Mr. Ch—— is always well dressed. His air and manners are gentlemanly.

He never speaks of business. He appears to be a man of independent fortune. But all this time the wolf has been turning his merchant friend to account. He has been seen with him on the Exchange, in the Baltic, and in other places " where merchants most do congregate," and those for whose benefit the wolf has thus paraded his acquaintance have, as they believe, ocular demonstration of what he has told them in private. They enter into transactions with him. Mr. Ch—— is playing a double game. And how does it all end? He has been introduced into the merchant's family. He has so worked on the sympathies of the ladies of the house that litanies of his praise are for ever ringing in the ears of the husband and father. Some fine morning the wolf leaves for Paris, and after his departure it is discovered that not only the honest merchant, but some of the sharpest of the City have been "let in" for sums which the wolf forgot to pay before starting, but which had been secured on stamped paper, crossed by his friends.

Mr. John M—— graduated at Oxford, and was afterwards called to the bar. He is well connected, is a member of a Pall Mall club, and was to be seen at the *soirées* of Lady P——, wife of a Cabinet Minister. His name used to appear in the *Court Journal* and *Morning Post* as a guest in high circles. He is personally acquainted with the editors of newspapers and writers on the press. There is, indeed, a misty belief amongst his acquaintances that he writes for the press himself, but of this nothing positive can be said. He has the *entré* into the different *ministères;* he is acquainted with the under secretary, or some respectable official, to whom he never fails to pay a weekly visit. But his visit is disinterested. He merely inquires after his friend's health, repeats an amusing anecdote, or retails a piece of astounding intelligence, learned at the house of one of his distinguished acquaintances. He is of no politics, and agrees in opinion with all whom he visits, no matter though they be diametrically opposed to each other.

Having performed his round of visits, he adjourns to his club, where he fares sumptuously, and where his dinner is washed down with sherry and champagne. He reprimands the servants for the slightest error, or what he deems inattention; he speaks sharply to the steward about the cooking, and threatens to report him to the committee, and in a loud voice pronounces the best Christopher's sherry to be "beastly stuff." In the same tone he speaks of his lands, his tenants, and his horses, and startles the whole room by the expression of his indignation if the second edition of the *Globe* or *Express* is not immediately laid before him. When he rises from table to transfer himself to the smoking-room, he moves with the ponderous dignity of a Roman cardinal or a Constantinopolitan archimandrite. Even the well-trained waiters cannot repress a semi-smile as they watch his movements. And this gentleman comes every day into our City, but people do not suspect him of doing business there. How could they? His is not the hasty step, nor is his the fixed, preoccupied eye of the laborious City toiler. Mr. John M—— strolls rather than walks through the City; he looks like some fair exotic amongst the wild flowers around—many of them running to seed, be it observed *par parenthèse*. When Mr. John M—— reaches a crossing, he does not hurry over with the eager haste of a hard-pressed man whose bills have arrived at maturity, and who has not the wherewithal to meet them. No. Mr. John M—— waits a convenient opportunity, and then, with a kind of friendly effrontery, he takes the arm of somebody who is crossing at the moment, graciously apologizing for the liberty, and saying that he is so short-sighted. Though, if Mr. John M—— is so short-sighted as not to be able to see his way across the street, I should like to know what is the use of the glass which seems to be permanently fixed round the orbit of his right eye.

But, as I have said, Mr. John M—— every day visits the City. In his easy way he saunters into the private office of some rich capitalist whose acquaintance he has taken pains to make at the club or in general society. He says a few words upon the topics of the day, or tells some piece of news that he has learned at the Foreign or Colonial Office, and so he makes his round of City visits. Mr. John M—— knows the hour at which his different City friends lunch, and it is due to his discrimination to say that he is well aware which amongst them keeps the best wine in the office, as he also never forgets which stockbroker smokes the best cigars after four o'clock.

But, reader, you will say Mr. John M—— leads a very idle life. Not at all. He does not practise at the bar. Very true. But he is engaged in trade. Yes, sir. *He trades in his acquaintances.* That is a branch of business of which you have possibly never heard. If Mr. John M—— makes his *periferia* regularly during ten months of the year in the City, it is because the City draws him with the same cords that it draws its professed *habitués*. I really do not know how to designate Mr. John M——'s actual occupations, unless I call him an INTRODUCER. That describes his real business. He introduces to each other men who seek a monetary advantage from one another's acquaintance. Does the promoter of a company want a few additional directors for his board, he speaks to Mr. John M——, who broaches the subject to some of his friends, sounds their feelings on the question, and ends in generalities if he see no hope of success, but comes to the point if the prospect be favourable. In the same way he introduces a stockbroker to a merchant or speculator. The introducer makes his terms beforehand, which are in proportion to the value of the services he renders. I paid Mr. John M—— about £1000 when I was bringing out my second steam company. It was more than I originally agreed to give, but there is something parasitical in the nature of the introducer. He clings to you with a wonderful tenacity. Even when he does not succeed in effecting what he originally undertook, he has often learned so much of your business that you cannot cut his acquaintance short, and to get rid of his officious interference you give him more than you had originally promised.

Though the introducee is generally the tool of the introducer, it sometimes happens that the latter is obliged to share his perquisites with the introducee. What do you say to this? Is not London the great *agora* of the world? Social position, a spotless name, a long-descended title, a staple monetary reputation, all find their price there. You may, if I dare so express myself, sell without bartering these commodities.

A man of high family connexions and of ancient lineage gives his *countenance* to a commercial concern, and for the halo thus thrown round the affair he reaps a respectable profit. What then? The transaction is all fair and above-board. His name gives a certain confidence to the public, and he is paid for his time.

It is sometimes very amusing to see how a man becomes the tool of an introducer. I was one day standing in the Exchange, talking with a friend. A gentleman, known to us both, stepped up, and begged permission to introduce Mr. Bodington, a highly respectable young man, who wished to speak with Mr. Price, my friend, about an affair that might be made advantageous to both. Mr. Bodington was a young man of prepossessing countenance, gentlemanly appearance, and good manners. I was about to retire, but Mr. Bodington begged me to remain, as the business might interest me too. After a few preliminary remarks, Mr. Bodington informed us that he was an importer of cigars, and requested Mr. Price, as a beginning of business, to take four boxes of regalias —a transaction of only about £10. Price, like myself, was astonished to find in such a swell a seller of cigars. In compliment to the gentleman who introduced him, he took the four boxes, quite unconscious that the introduction was in itself a "transaction." Mr. Bodington's mode of proceeding was this: He sought to do business with wealthy City men, supplied them with the best cigars, and rarely troubled them for money. We may suppose that his profits equalled his forbearance, and hence the importance to him of an introducer. Mr. Bodington carried on a profitable and honourable trade, and was a kind-hearted man. *A propos* of his kind-heartedness, I shall narrate a circumstance, sufficient, I think, to reconcile the most cigar-detesting wife to the tobacconist. I know a great Indian merchant who became bankrupt some years ago. His liabilities amounted to some hundreds of thousands of pounds sterling. He owed Bodington £165 for cigars. Little could the proud merchant have dreamed, in the days of his prosperity, that the humble tobacconist would one day become his benefactor. And yet, so it was. Not only was Bodington the first to sign the bankrupt's compromise, but he lent him, on his personal security, £1500 to recommence business. It was the mouse liberating the lion from the net. That merchant is again a rich man; he returned liberally the money lent to him by Mr. Bodington, but I question whether his pride would not be hurt were it known that he is indebted to a humble trader for the materials with which he reared the second temple of his greatness.

But I have strangely wandered from my subject. I was about to show how some men make great profits out of a small capital, and how others work on and make a fine appearance without any capital at all. I have seen empty-pocketed speculators tempt rich sanguine-minded men to their ruin, and I have seen the latter, when ruined themselves, become in their turn tempters.

Mr. Peter P—— for many years had carried on a safe trade. He was one of those worthy, reliable men who are chosen as umpires in cases of arbitration, and who are appointed trustees to the property of widows and orphans. One afternoon, Mr. B——, a bill discounter, stepped into his office, and coolly asked him whether he wanted money. The merchant stared. "Because," said the other, "I have £3000 which I wish to invest at 5 per cent. If you like, you can have it for six months or a year, put it in your business, and I will take your bills for six months. You can keep the money longer than a year if you wish, and I shall always renew by adding the interest, 5 per cent." Money was very cheap in the City at the time. The merchant knew that he could make 10 per cent. on this fresh capital thrown into his business. Money is a temptation that few men can resist. He consented. His clerk drew and endorsed the bills. They were duly accepted. Mr. B—— went off immediately to Lombard Street, where he discounted the bills without endorsing them. Having been afterwards mixed up in the affair, I had opportunities of learning all the particulars. No sooner had the bill-broker discounted Mr. Peter P——'s bills at 3 per cent., than he went to another highly respectable trader and made a proposition similar to that which he had made to my poor friend, and with like success. Now, we must not suppose that Mr. B—— was the real owner of the £3000 which he had lent Mr. Peter P——. No. The £3000 were part of the product of bills that he had previously "melted" at the "Corner House," without his indorsement, for several unstable firms that entrusted him with such transactions. He did not immediately pay back all the money to his patrons. In such cases it is easy to frame a plausible excuse—give them a cheque on account. He expected to "do" them upon good terms within a few days. In such case he was the debtor to several unstable merchants to the aggregate amount of between £10,000 and £15,000. He will pay them this balance when their discounted bills are paid off. When Peter P——'s first bills fell due—viz., at the end of six months—money had risen, and Peter asked for a renewal. He was informed by Mr. B—— that the discounter no longer held the bills—that they were in the hands of a third party. What was to be done? Peter could not meet the bills. He had goods, but these he could not so summarily dispose of, except at a great loss. After much negotiation, Mr. B—— said that the holder was willing to renew the £2500 for three months longer, at 1 per cent. above the bank rate of interest, and with 5 per cent. commission on the amount. Poor Peter P—— was in despair, but was obliged to accept the offered terms. Mr. B—— advised the Lombard Street firm to renew at the bank rate, and with a commission of 2 per cent., Peter's bills, as well as all the other bills of the shaky merchants, and wait until the crisis passed. As the melting-houses do not like to hold dishonoured bills, the renewal for three months was made as advised, and the broker pocketed 11 per cent. on money that was never his.

But Peter P——'s troubles were not at an end. When the renewed bills fell due, he was obliged to pay still heavier interest than before, and higher commission; and so things went on until he was enveloped in a mesh-work from which he could not break; his struggles were vain. At length, in his desperation, he left the bills unpaid, and threatened his creditors to become a bankrupt. Then the last turn was given to the vice in which he was held. The bills were put into the hands of a solicitor, who first wrote a letter, then issued a writ, and poor Peter, who was still solvent, and anxious to save his credit, gave, as security for the bill debts, goods for part of which he had not yet paid. But the strangest circumstance in the case was, that he gave an additional lien on these same goods to Mr. B——, whom he all along believed to be responsible for the amount of the bills. It was at this stage of the proceedings I was called in. I thought that what Peter P—— had made such efforts to stave off was inevitable. I suggested Basinghall Street. The thought of becoming a bankrupt was to him, as to every honest man, hateful. He still carried on the death-struggle. And how did it end? Driven to the last extremity, he still rejected the idea of bankruptcy. He compromised with his creditors. Mr. B—— was his assignee. All his goods were sold, and after the creditors had received their dividend, Mr. B—— pocketed the rest, as well as the greatest part of the balance, £15,000, of the shaky houses, most of which stopped payment, as a compensation for his trouble.

You will tell me that Peter P—— was a fool to accept the offer of the broker in the first instance. To be sure he was. But do we not think the whole English monetary system anomalous, and calculated to tempt men to their ruin? I know, and you must know, reader, many whose name, ten or twelve years ago, would be sneered at, and whose name is now good for thousands. And how did they rise? That is their secret. A decade of years since, and they were hard-working paupers, and now they are the table companions of millionaires. In fact, such men are like aëronauts; if they can only get a sufficient quantity of gas lighter than the atmospheric air, they will rise and rise, and continue to float, until the gas by some mischance escapes, or a rent comes in the bag, and down they come.

It was in the year 1851 that I frequently met, at the *Café de l'Europe*, Haymarket, amongst dramatic people, and in the company of the late A—— S——, a young man, of whose avocation in life I knew nothing. He could be seen often in the neighbourhood of the Haymarket Theatre, and hanging about gaming-houses. I lost sight of him for several years. Circumstances had made it advisable for him to cross the Channel. In 1860, I one day met my old acquaintance in a billbroker's office. Mr. Manley—I now learned his name —was pleading hard to get a small bill discounted. I soon after heard that he had become secretary to an extensive shipowner. Through this gentleman, Mr. Manley obtained an entrance into the establishment of Messrs. Overend, Gurney, and Co., of Lombard Street. There he breathed a congenial atmosphere, and throve apace. The intimate knowledge he possessed of his late employer's affairs was placed at the service of his new masters, and gold chinked in his purse. Initiation into the Samothracian mysteries of old commenced by placing the neophyte on a throne. Mr. Manley's occupation of the high stool in the great office was the first step on the ladder by which he ascended to a seat in the House of Commons, where he now sits a dumb lawmaker.

A petty wine merchant named Stanley, discovered that discounting bills is more profitable than selling sherry. In 1859 he established a small discounting company. As he was a man whose business capabilities I recognized, I invested £1200 in the concern. This I did not alone because I believed the affair would prosper, but because I really wished to assist the promoter of the scheme. However, the project came to grief. Mr. Stanley had extended his patronage in a very marked manner to members of the leather trade. He discounted bills to the amount of £70,000 for great leather merchants, who afterwards failed for £500,000, paying a dividend of one shilling in the pound. Disappointed in one speculation, we were obliged to wind up. We—that is the shareholders—lost a considerable sum, and were obliged to pay Mr. Stanley a good price for giving up the management. Mr. Stanley, at least, had lost nothing. But his name was not sweet-sounding, and he thought it would be well to change it. He became Mr. Jackson. He continued to knock about the streets of the City, like the thousands who crowd those great thoroughfares, when towards the close of 1863, when the mania for companies was at its height, he assembled a dozen manageable men, and brought out a financial company. It was a wonderful company, for not only did it go on prosperously, but, like Paley's celebrated hypothetical watch, it brought forth young companies. Five times in the course of one year did the prolific mother present to the world a baby company. But these babies quickly became strong and vigorous, and were able to go alone. Still it must be admitted that the mother laid heavy contributions on the children. In fact she took away half their food, and cramming a great deal of their property under her voluminous crinoline, astonished her admirers by her portly appearance. But occasionally taking from beneath her crinoline some of her children's property, she presented it to her supporters, just as a *bonâ fide* company gives dividends to shareholders.

I have several times attempted to congratulate my old friend in person, but have never been able to obtain an audience. Behind the panels of his polished mahogany door I found a tall liveried footman, who, in reply to my inquiries, informed me that his master was in court. I returned again and again, and always received the same answer. I at length asked to what court Mr. Stanley, or rather Mr. Jackson had gone. The official replied with a gracious smile: "Sir, we call our board meetings 'courts.'" That was sufficient.

If I had been misled, it was for want of inquiring at first. We live in a free country, where everybody has a right to do what he pleases with his own, and call it by what name he likes.

I said I would wait until Mr. Jackson should be at leisure. I was invited to walk in, and soon found that I was not alone " doing ante-chamber," to the *ci-devant* wine merchant. Would you believe it? I saw waiting in that ante-room one of the great Cohens. I know not how long he had been there before I arrived, but I know he remained there a " large half hour," as the French say, after I entered. The man of ancient greatness waiting on the man of recent greatness, looked rather awkward. He moved about occasionally —looked out of window—all the while rubbing his eye-glass with his cambric handkerchief. At length he was admitted to the audience-room.

Seeing these things, I began to surmise what kind of reception I might expect. I recalled past times, and said within myself that I had some claims on Mr. Stanley. I remembered that I had formerly put large business in his way, and had done him some good turns.

At length I was admitted to the presence of the great man. He gave me a cordial reception, hoped to see me again hold a high place in the commercial world, but he could not assist me to launch my little steam company, being himself engaged in bringing out a very large one. I further learned that Mr. Stanley's fleet was to be composed of the steamers formerly belonging to Messrs. Overend, Gurney, and Co., of Lombard Street. What a change! I remembered how one day, as I was stepping into the great " melting-house "—I was at that time on good terms with Messrs. Overend, Gurney, and Co.—I met Mr. Stanley coming out, his hand full of bills, with tears in his eyes. "That d—— Ch——," he said, "is a nasty fellow; he would not do anything for me." And now Messrs. Overend, Gurney, and Co.'s "melting-house" has melted away, self-consumed, but Mr. Stanley had become a representative of the people—a nobleman—and only waits another opportunity to dazzle the world. Will you not, reader, admit that the acquisition of money is the end and aim of modern civilization?

Perhaps you will ask me what has brought about this paradoxical state of things. In reply, I shall say "a commercial revolution." Commercial, like political revolutions, occur once or twice in every century. Commercial revolutions' generally occur when money accumulates in the possession of traders; when certain branches of business fall exclusively into the hands of a few monopolists; when articles, once esteemed luxuries, become necessaries; when an entirely new article is brought into the market in sufficient quantities to vie with the old; or when an article well known, but the product of a remote clime, is suddenly brought, by improvement in the mode of transport, into the market, and made to compete successfully with the old standard commodities. Then may be seen commercial Robespierres and Marats springing up in every branch of trade, wresting power from ancient royalty, trampling down the once privileged legitimists, and building on the ruins thus made the fabric of their own greatness.

Let me give you some examples:—Some fifty years ago cotton from Smyrna and India was a Pactolus that poured gold into the coffers of a few men who had monopolized the trade. About that time the cotton of South America was introduced into the English market. A few speculators in Liverpool and Manchester seized the opportunity, took possession of the new vein of commerce thus opened, and the cotton trade with the East was, so to speak, stamped out. Twenty years ago, how many Greek merchants had made fortunes, and how many were growing rich, by the corn trade of the Levant; but after the Crimean War, and since the late American War, the wheat and maize of America have taken the place, in the English markets, of the Euxine corn, and many a wealthy Glaucus has been obliged to return to the Levant or become a broker.

I must, however, observe that the commercial revolution of 1863, which suddenly converted the petty wine merchant into a wealthy man, and gave the toss-up that made the penny gambler of the *Café de l'Europe* a representative of the people, such a revolution is, I must say, unprecedented in the annals of commerce. That revolution did not affect one branch of trade alone; it was universal, and resulted from the combined action of those causes which I have already named as the secret springs of commercial revolutions. Money had accumulated on every side in the City; telegraphic and steam communication supplied an hourly stimulant, which, in previous times, could not be offered to the fever of speculation. Then there was the mania that seized England of making large loans to foreign countries, and more particularly to rotten Governments that paid large dividends by the ingenious device of getting a fresh loan. There was besides that extraordinary amalgamation, which you must remember, of West-end swells, with hard-working City men.

But that is not all. One of the partners of an eminent firm having dipped himself somewhat too deeply into debt, conceived the idea of turning the private concern into a limited company. The other partners chimed in. They were equally glad to get rid of their responsibilities. The firm sold their goodwill. Now the goodwill, *entre nous*, was substantially a consignment of insolvency. The *grand livre* of the public was debited with the enormous debts contracted by the young scamp on the turf and the Stock Exchange. There were other items, such as a theatre rented for the convenience of his favourite actresses, and a hotel started for a *ci-devant* mistress. There had been a large sum expended in making a provision for five illegitimate children.

You must not suppose that our commercial revolutions do not bring forth heroes. He who stands his ground and holds his own is a good general. He who succeeds in erecting a fabric, more or less permanent, on the *débris* around, is reputed a conqueror. I have seen heroes that achieved great things, but whose "vaulting ambition" at some unlucky moment "o'erleaped itself;" 't was then the hero fell, and was borne off the field to the great hospital of Basinghall Street. In such conflicts the number of victims is large, that of conquerors small. In the general confusion the motto of each is *sauve qui peut*, and self, self alone, is in the ascendant. Then the voice of honour, of feeling, of humanity itself is silenced. No one gives or expects quarter. Friendships, family ties, everything is forgotten—lost in the sense of personal danger. What a spectacle does the human heart exhibit then, a prey to the pitiless ferocity of *philargyric* fury!

I have said that some of our heroes are carried off to Basinghall Street hospital. The wounded sufferers of commercial revolutions afford abundant occupation to the surgeon, physician, and coroner. These " professionals " in our world are known as solicitors, accountants, and liquidators. The wounds and sufferings of the victims are an inexhaustible source of profit to these servants of the Adrastian Nemesis.

Such a revolution as I have described generally lasts some years. A few more must elapse before the great heart of the commercial body again makes healthy music. Leaving the City restored to its normal state, let us look round on the suburban districts, where our merchant princes make homes for their families. We shall find great changes in these places. The earthquake that shakes the City generally leaves more permanent traces of its action at remote points than at the centre. These changes are in persons, not in things. The earth in these localities still presents the same appearance that it wore before the occurrence of the great catastrophe. The houses and trees, the beautifully laid-out gardens, and the stately entrances to the noble mansions still stand to each other at the same relative distances that they ever did. But, alas! alas! the voices that now issue orders in these houses are not those of the former owners. The eyes that look with satisfaction on the handsomely furnished rooms and spacious lawns are not those of him who formerly looked on them and said in his heart—"This is all mine."

I knew a great capitalist who, in 1863, was living at Kensington. After the great crisis of that year, I saw the man who had accepted and honoured bills to the amount of millions of pounds sterling, living in an obscure house in the City Road; and I saw a commercial *sans culotte* become the legal owner of the fine house at Queen's Gate. What tears the proud Mrs. Polycrates and the fashionable Misses Polycrates shed when the auctioneer coolly knocked down to a five pound bidder the elegant Dresden china that had cost £100; and the Rubens and the Spagniolotti, for which Polycrates had paid £500, went for £10, being then discovered to be, not originals, but copies. The beautiful horses and carriages passed into the possession of the fortunate *sans culotte*. Mr. Briscott—that is the name of the lucky man—had passed all the years of his married life in a little cottage at Tottenham. There he had brought up a large family of sons and daughters. He was a petty bill discounter, and made his great *coup* in the Stock Exchange and in cotton during the crisis.

Were I to go into some of the details which came under my own observation in the case of these two families, you would certainly laugh, as I have done, even whilst I sincerely commiserated the fallen family. Easter was drawing nigh. The Briscotts were fully established in their grand mansion. Passion week was to them gay as the Roman carnival. The mother did not become familiarized as easily as the father and daughters with her newly acquired luxuries. I know it for a fact, that the first time she drove out in her carriage she begged her husband to keep the blinds close, as she could not get rid of the idea that everybody was staring at her. Poor Mrs. Briscott felt as awkward in her fine coach as though she were mounted on a Smyrniote camel, and paraded through an itinerant fair.

But Mrs. Briscott was not a woman to stumble at trifles. She was determined to enjoy the good things that had fallen to her lot. Good Friday arrived, and Mrs. Briscott, no longer occupied with the care of hot-cross buns and salt fish, with an accompaniment of parsnips, determined to attend a fashionable church in Belgravia. When she arrived, nearly all the seats were occupied. The preacher was of High Church principles, very eloquent, and much beloved by the titled ladies who formed the majority of the congregation. Mrs. Briscott slipped two sovereigns into the pew-opener's hand. She and her daughters were soon seated where their showy toilettes might be most fully displayed. Mrs. Briscott in this way did a pious and recreative piece of business on Good Friday.

How did the Polycrates family pass the day? Being a holiday in the City, Mr. Polycrates proposed to his wife and daughters to take a trip to Greenwich. The ladies had been but little out of doors since their reverse of fortune. To do them justice, they had endeavoured to accommodate themselves to circumstances. The daughters had taken up their needles—instruments they were little accustomed to touch—and busied themselves in the thousand alterations and repairs which the wardrobes of struggling people require. The mother had taken upon herself the superintendence of the one servant employed to do the household work. It is not to be supposed that in that house in the City Road there were no tears shed, no sighs fetched, no complaints made. There were, many and many. Tempers were not always as smooth as they used to be at Queen's Gate, nor were laughs so frequent and merry. But all did their best, and the toil-pressed husband and father was received with a cheerful welcome when he returned to his humble home of an evening.

When Mr. Polycrates proposed taking his family to Greenwich, he thought only of giving his wife and daughters a little fresh air; but he forgot the accessories that must accompany their trip. The ladies knew nothing of these inevitable inconveniences, and though they were at first disgusted by the very name of Greenwich, they afterwards took a more philosophical view of the affair, and thinking it as well to conform to their altered circumstances, set out for the London Bridge Station. The young ladies were horrified at the rough crowd, at their loud laughs and coarse jests. They were disgusted at the predominating odour of *zythos* (ale). In the general confusion the members of the Polycrates family were separated, and each stepped into a different carriage. Mrs. Polycrates found herself seated beside an unreduced Banting, the magnitude of whose figure gave him a certain air of respectability. A poor man could not have attained such dimensions. The portly gentleman tried to make himself agreeable to the full-blown matron. He began by talking of the weather, and then launched into the history of several of the aristocracy, with whose genealogies he exhibited a great familiarity. It is evident that Mrs. Polycrates has made a conquest. The stout gentleman gives her to understand that he is a bachelor, and in good circumstances. Seeing the lady alone, he proposes that when the train stops they shall take a walk in the park, and have tea or coffee in some of the pretty garden-houses. Mrs. P. declines with a smile. All this time, the lady, as she afterwards told us, laughing heartily, thought that her fat acquaintance was a rich City man closely connected with the aristocracy.

The eldest Miss Polycrates found herself sitting opposite to a very handsome gentlemanly young man, a clerk in a bank. He cannot help looking at the young lady, but he is too timid to speak. This silent homage is not displeasing to her to whom it is offered. When the train stopped, the young man jumped out of the carriage, and presented his hand to assist the lady to alight.

The other daughter was not so fortunate. She chanced to sit next a blacksmith, dressed in holiday finery. A purple waistcoat; a red necktie; a white shirt, with three brilliant glass studs; a steel watch-guard, and a pair of plaid trousers, formed a toilette which, with some potations of the plebeian nectar, had put the blacksmith on such very good terms with himself, that he attempted to be facetious with the lady. But Miss Amy Polycrates was high-spirited and resolute. She had sat at first with her silver-headed smelling-bottle in her hand, looking ineffable things; but, when the blacksmith tried to jest, she told him in a determined tone that she would call the guard.

The father was the most fortunate of the party. He was seated between two pretty young milliners. They were full of lively chat, and seemed determined to enjoy their holiday.

When the train had reached the station, and the scattered members of the Polycrates family were again united, Mr. Polycrates recognized, in his wife's stout travelling companion, a porter in one of the municipal establishments in the City. The Corporation official, to whom Mr. Polycrates' features were familiar, was overwhelmed with confusion, and hurried away as fast as the load of fleshy frailty with which he was burdened would permit. The banker's clerk followed the party unobserved, at a distance, anxious to discover the name and address of the charming young lady.

The amusements at Greenwich afforded no pleasure to the Polycrates family. The kiss-in-the-ring disgusted the daughters; the donkey-riding depressed the spirits of the father, who thought of his beautiful steeds, Blondin and Leotard; the English *dervishes*—those ignorant and fanatic open-air preachers—addressing crowds of half-intoxicated people, and the sight of the sleepers on the grass, threw Mrs. Polycrates into a melancholy mood. The whole party left Greenwich three hours before the *panegyris* concluded.

We left Mrs. Briscott at church on Good Friday. I should be doing that lady a very great injustice, if I allowed you to remain under the impression that she never became acclimatized in the atmosphere of Queen's Gate; she did, and ruled there as absolutely as she had always done in her Tottenham cottage. Once thoroughly initiated into her new sphere, her real character appears in its strongest colours. She is tyrannical to her servants; she is stingy to the individual poor, but ostentatiously liberal in her subscriptions to public charities. She is proud and narrow-hearted towards her indigent relatives; she is *gauche* in her movements, and quarrels with her husband for the most trifling cause. She frequently reminds him that, though she was his cook before she became his wife, still it was owing to her good advice, and her economical management at Tottenham, that he came to be in a position to make his money. Mr. Briscott is now a Member of Parliament, but his wife is greatly annoyed at his remaining late at night in the House. She is also angry at his being a dumb legislator; she wishes him to make a speech. She longs to see his name in the *Times*, under the head, "IMPERIAL PARLIAMENT." If she were in his place, she tells him, she would let the House hear as well as see her; and, no doubt, she would.

Reader, you perhaps think that I draw upon my imagination; if so, you are mistaken. I could point you out, within the circle of my own acquaintance, ten families who have undergone changes similar to those I have described. I have seen wealthy men fall with a sudden crash and never rise again. I have seen petty brokers, insignificant traders—adventurers that used to hang about my office trying to make five pounds' commission, or hoping to induce me to promote some imaginary scheme of theirs, or enter into transactions with them—I have since seen these men, I say, rise into sudden wealth, and I see them now occupying distinguished positions in society.

You will naturally ask me to account for the fall of the one and the rise of the other.

Was the first an idle and extravagant, or an unfortunate man? Was the second a commercial genius, or simply a daring, fortunate speculator? In ordinary times people do not fall or rise so rapidly. The clever and industrious decidedly rise, but cannot, within twelve months, rise from nothing to be worth £150,000 and obtain a seat in Parliament. Such extraordinary *phenomena* are the results, as I said before, of commercial revolutions. The commercial revolution of 1863-65 was one of three stages. During its existence nearly all England was thrown into a state of frenzy. The prudent rich man then lost his head, when the desperate pauper found his own. The latter rushed to the field, having nothing to lose, but where he could most easily gain a great deal.

The mania for bringing out limited companies was the first stage of that revolution. A dense fog then covered the atmosphere of the City. You could no longer distinguish the good shares and bills of exchange from the bad, because many of these worthless concerns were under the covers of the names of noble dukes, earls, and lords; gallant generals, colonels, admirals, commodores, and captains; and eminent insolvent merchants and bankers. The gold of the public accumulated in these institutions, which seemed to promise a new *golden era* to commerce. The capitalists naturally followed, as they always do, the public spirit. They bought the shares, and exchanged their sterling gold for these fine printed, decorated, and sealed papers. They were saddled, besides, with a tremendous responsibility, which they little thought of then.

The second stage of the revolution begins when masked fire brigades poured in to arrest this *paroxysmos* of turning insolvent businesses into "limited companies." These new comers showed themselves determined to pull down even the soundest concerns. To them alone the panic is due. They first gave the cry, "*Sauve qui peut!*" The shareholders showed no courage, unanimity, or determination to protect their companies. Alarmed, they fled before these masked alarmists, and left them abundant spoil.

The third stage of this revolution is when the law despatched its servants, the *liquidators*, to take care of the remainder of the shareholders' property, with full permission to take care of themselves, too, by charging a commission on the debts of the companies. We all know the result of this system. Companies that had debts to the amount of £5000 found those debts increased to ten and twenty times that sum. Therefore, reader, it was not extravagance or incapacity that brought Mr. Polycrates to his present position, nor commercial talent that raised Mr. Briscott to his. Mr. Polycrates was a *bonâ fide* large shareholder and capitalist: Mr. Briscott was a promoter, fire brigadier, and polynomial liquidator.

It will be a long time before we again see business and confidence restored in the *City of London*.

APPENDIX.

No. 1.

Messrs. Overend, Gurney, and Co.'s compliments to Mr. S. Xenos, and beg to inform him that they have given Messrs. Coventry, Sheppard, and Co. orders to sell the whole of the wheat per Admiral Kanaris, and request that Mr. Xenos will withdraw his orders from any other parties whom he has instructed to sell.

Lombard Street,
    31 *Dec.*, 1860.

No. 2.

London, January 12, 1861.

Messrs. Overend, Gurney, and Co.
    Gentlemen,

I have to request that you will advance me this day £700 (say seven hundred pounds) for the purpose of meeting certain engagements of the Greek and Oriental Steam Navigation Company, and in consideration of your so doing, I hereby engage that, should you think it desirable hereafter to form fresh combination of capitalists for the purpose of carrying on the said company, that I will give you every assistance in my power.

    I am,
        Gentlemen,
            Your obedient servant.

No. 3.

Messrs. Overend, Gurney, and Co.'s compliments to Mr. S. Xenos, and beg to say that they understand that he has an offer from the ship-builders in the North for the two vessels, the Palikari and Odesseus Androutzos, and they request that he will accept the same at once, otherwise he must expect no further consideration from them in any way.

Lombard Street,
  21 *Feb.*, 1861.

No. 4.

Messrs. Overend, Gurney, and Co.'s compliments to Mr. S. Xenos, and beg to request that, in chartering the steamers, he will insert a clause giving him liberty to cancel the agreement in the event of the ship being sold previous to her departure from London.

Lombard Street,
  21 *Feb.*, 1861.

19, London Street.

No. 5.

Messrs. Overend, Gurney, and Co.'s compliments to Mr. S. Xenos, and will be obliged by his furnishing them with the annexed particulars of each ship in his fleet, which are wanted for the purpose of forming a company. O., G., and Co. would like to have them to-day, if possible.

Lombard Street,
  25 *Feb.*, 1861.

> Cargo, dead weight.
> Nature of engines.
> Name of maker, &c.
> Consumption of coals *per diem*.
> Quantity of coals bunkers will hold.
> Number of passengers.
> Price, minimum.

No. 6.

Messrs. Overend, Gurney, and Co.'s compliments to Mr. S. Xenos, and beg to say that, till the letter of Mr. Chapman's is signed, they cannot make any further advances, and request he will do so at once, and forward it to them.

Lombard Street,
    19 *Jan.*, 1861.

19, London Street,
    Fenchurch Street, E.C.

No. 7.

ΑΓΓΕΛΙΑ.

Θέλει δημοσιευθῆ ἐν Λονδίνῳ εἰς τὴν καθομιλουμένην τὴν 1ην Ἰουλίου τοῦ παρόντος ἔτους ὁ ΒΡΕΤΤΑΝΙΚΟΣ ΑΣΤΗΡ, ἐφημερὶς Πολιτική, Δικαστική, Ἐμπορική, Φιλολογική, Ποικίλη καὶ μετὰ εἰκονογραφιῶν.

Ἕκαστον φύλλον της, μεγέθους τοῦ σχήματος τῶν ἐφημερίδων τῶν Ἀθηνῶν, θέλει σύγκειται ἐξ 24 σελίδων μετὰ τεσσάρων στηλῶν καὶ πεπυκνωμένης ὕλης.

Ὁ Βρεττανικὸς Ἀστὴρ θὰ ἐξέρχηται τῶν πιεστηρίων ἐκάστην Πέμπτην, ὥστε νὰ στέλληται εἰς τὴν Ἀνατολὴν τακτικῶς διὰ τοῦ Γαλλικοῦ ἀτμοκινήτου.

Τὸ Πολιτικὸν μέρος θέλει διαιρεῖσθαι εἰς τρία διακεκριμένα μέρη:

ά.) Τὰ κύρια ἄρθρα, ἀφορῶντα τὸν πολιτικὸν ῥοῦν τοῦ πολιτισμένου κόσμου.

β'.) Τὰ νομοσχέδια ἐκεῖνα καὶ τὰς ὁμυλίας καὶ συζητήσεις τῶν ἀγγλικῶν βουλῶν, τῶν βουλῶν τῆς Ἀμερικῆς, Γαλλίας, καὶ ἐν γένει ὅλων τῶν ἐλευθέρων Ἐθνῶν, τὰ δυνάμενα νὰ χρησιμεύσωσιν ὡς ὑπογραμμὸς πρὸς ἡμᾶς διὰ τὴν κρείττωνα ἐφαρμογὴν τοῦ Συντάγματός μας, πρὸς φωτισμὸν τοῦ Ἔθνους.

γ'.) Τὰς εἰδήσεις, καὶ τὰ τηλεγραφήματα ὅλων τῶν μερῶν τῆς Εὐρώπης, Ἀμερικῆς, Ἰνδιῶν, Κίνας, Αὐστραλίας, ἢ μᾶλλον τί λαμβάνει χώραν εἰς τὸ λοιπὸν τῆς ὑδρογείου σφαίρας.

Τὸ Λονδῖνον εἶναι ἡ Ἑστία τῶν εἰδήσεων· καὶ ὅστις ἀναγινώσκει τὸν Ἀγγλικὸν Χρόνον καὶ τὰς ἄλλας Ἀγγλικὰς ἐφημερίδας, μόνος ἐκεῖνος δύναται νὰ ἐννοήσῃ, ὁποία τερπνὴ καὶ ἀπέραντος σχολὴ εἶναι τὸ μέρος τοῦτο.

Τὸ Δικαστικὸν μέρος εἶναι μὰ μυθιστορήματα τῆς ἐφημερίδος, ἢ μᾶλλον ἡ ἀντιγραθὴ τῶν Κοινωνιῶν τούτων. Ἡ Δίκη τοῦ Ἐναγοῦς

Ἰατροῦ δεικνύει καὶ τὸ ὗφος, καὶ τὴν περιγραφὴν, καὶ τὸν τρόπον τοῦ διεξέρχεσθαι μίαν δίκην. Θέλει δὲ διαιρεῖσθαι εἰς τρία ἐπίσης μέρη:

ά.) Τὰ καθημερινὰ πταίσματα δικαζόμενα ὑπὸ τῶν λεγομένων, Κριτῶν, μέρος διασκεδαστικὸν, καὶ συγχρόνως σπουδαῖον, διότι ἐν αὐτῷ θὰ κατοπτρίζηται ὁ βίος τοῦ ὅχλου, καὶ τὸ ἀστυνομικὸν σύστημα ὑπὲρ τῆς ἐξασφαλίσεως τῆς τιμῆς, καὶ τῶν ὑπαρχόντων ἕκαστον πολίτου.

β'.) Τὰ κακουργήματα, ἀνάλυσις ἐπιστημονικὴ τῆς πράξεως, καὶ περιγραφὴ θαυμασία τοῦ ὁρκωτικοῦ συστήματος, τῆς ἰσχύος τοῦ νόμου, καὶ τῶν διακεκριμένων ῥητόρων τῆς ἡμέρας.

γ'.) Αἱ δίκαι τῶν διαζυγίων, ἔτι περιεργοτέρα ἀνάγνωσις, πλὴν καὶ πιστοτέρα καὶ ζωηροτέρα περιγραφὴ τοῦ ἀγγλικοῦ χαρακτῆρος, καὶ τῶν βάσεων τοῦ συζυγικοῦ βίου, ἐξ ὧν ἐξήρτηται ἡ ἠθικότης καὶ εὐημερία ὁλοκλήρου τῆς Κοινωνίας.

Τὸ Φιλολογικὸν μέρος θέλει περιέχει, βιογραφίας, περιγραφὰς, ταξείδια, βοτανικὰς καὶ ζωολογικὰς ἐκθέσεις, ὡραίας τέχνας, περιηγήσεις, τὴν ἐπιθεώρησιν τῶν νέων συγγραμμάτων, καὶ τὴν θεατρικὴν καὶ δραματικὴν ἐπιθεώρησιν τοῦ Λονδίνου καὶ τῶν Παρισίων.

Εἰς τὸ Ἐμπορικὸν μέρος ὁ ἔμπορος θέλει εὑρίσκει μακρὰ τιμολόγια τῶν διαφόρων εἰδῶν, τὴν ἐμπορικὴν ἐπιθεώρησιν τῶν ἀγορῶν Ἀμερικῆς, Ἰνδιῶν, Αὐστραλίας, καὶ τῆς Εὐρώπης· τὰς παρακαταθήκας καὶ προβλέψεις τῶν ἐμπορευμάτων, καὶ πᾶν ὅ,τι ἀφορᾷ τὴν ναυτιλίαν, τὸ ἐμπόριον καὶ τὴν βιομηχανίαν, καὶ τέλος τὸ χρηματιστήριον καὶ τὰ ἐθνικὰ δάνεια, καὶ τὰ, συναλλαγὰς τῆς ἑβδομάδος, ἅπερ εἶναι τὸ θερμόμετρον, τοῦ ἐμπορίου.

Ὑπὸ τὴν λέξιν Ποικίλα, περιλαμβάνομεν τὸ ἐπιστημονικὸν μέρος, δηλ. τὰς ἀνακαλύψεις τῶν τεχνῶν, τὰς τελειοποιήσεις τούτων, τὰς τελειοποιήσεις τῶν ἐργαλείων, τὰς νέας ἐφευρέσεις, τ' ἀνέκδοτα τῆς ἡμέρας, τὰ βιομηχανικὰ ἔργα, τὰ ἐργόχειρα τῶν σχολείων, καὶ τέλος πλῆθος ἄλλων ἀορίστων ἀντικειμένων, ἅτινα σχηματίζουν τοὺς μεγάλους τούτους κόσμους.

Ὁ σκοπὸς τοῦ Βρεττανικοῦ Ἀστέρος δὲν ἀποβλέπει εἰς ἄλλο, εἰμὴ νὰ γείνῃ ὡς ὀχετὸς τῶν φώτων τῆς Δυτικῆς Εὐρώπης εἰς τὴν Ἀνατολὴν, καὶ ἐξ ἄλλου νὰ καταστήσῃ ταύτην αὖθις γνωστὴν εἰς τὴν Εὐρώπην.

Ἐν παρόδῳ δ' εἰδοποιοῦμεν, ὅτι πρὸς τακτικὴν δημοσίευσιν τῆς εἰρημένης Ἐφημερίδος, ἤδη τὸ πρώτων συσταίνεται ἑλληνικὸν τυπογραφεῖον ἐν Λονδίνῳ, καὶ ὅτι δέκα εὐφυεῖς μεταφρασταὶ, εἰδήμων ἕκαστος εἰς τὸν κλάδον του, θέλουν ἐργάζεσθαι ἐπὶ ταύτης.

Ἕκαστον φύλλον τῆς Ἐφημερίδος ταύτης, ἐκτὸς τῆς πεπυκνωμένης ὕλης αὐτῆς, θέλει περιέχει καὶ 15—25 εἰκονογραφίας μεγάλας καὶ μικρὰς ὡς ἐν τῷ τόμῳ τούτῳ.

Ἡ ἐτησία συνδρομὴ τῶν 1248 σελ. τῆς εἰρημένης ἐφημερίδος προσδιωρίσθη αντί τριών Γκυνέων, ή τριών λιρών και τριών σελινίων κατ' έτος, προπληρωτέων ευθύς μετά την παραλαβήν του πρώτουφύλλου.

Αί συνδρομαί γίνονται μόνον ετησίως παρ' άπασι τοις πρακτορείοις της 'Ελληνικής και 'Ανατολικής 'Ατμοπλοϊκής 'Εταιρίας, ή δι' επιστολής προπληρωτέας προς τον εν Λονδίνω κύριον EDMONS, 19, London Street, Fenchurch Street.

Η ΕΛΛΗΝΙΚΗ ΚΑΙ ΑΝΑΤΟΛΙΚΗ ΑΤΜΟΠΛΟΙΚΗ ΕΤΑΙΡΙΑ ΕΓΓΥΑΤΑΙ ΤΗΝ ΕΚΔΟΣΙΝ ΤΗΣ ΕΦΗΜΕΡΙΔΟΣ ΤΟΥΛΑΧΙΣΤΟΝ ΕΠΙ ΔΥΟ ΕΤΗ.

Δημοσιεύοντες την παρούσαν Αγγελίαν παρακαλούμεν τον φιλαναγνώστην να μη την εκλάβη ως δίσκον ελέους, όπως ρίψη τον έρανόν του. Τουναντίον εννοούμεν να τω παράσχωμεν πράγμά, πολύ ανωτέρας αξίας της πληρωμής του. Εννοούμεν να φέρωμεν εις την αγοράν ύλην, την οποίαν ο ίδιος αυθορμήτως ν' αγοράση δι' εαυτόν και τα τέκνα του, άνευ παρακλήσεων άλλων συστάσεων, ή φιλικών μέσων.

Είθε ο Παντοδύναμος να κατευοδώση τα διαβήματά μας! και είθε οι απανταχού αληθείς Έλληνες να κατανοήσωσι το έργον μας, και εννοήσωσι ότι δεν πρόκειται περί ατομικών υλικών συμφερόντων (τουναντίον, ελήφθη προ οφθαλμών μεγίστη χρηματική ζημία διά τα πρώτα έτη), αλλ' υπέρ των ηθικών και συμφερόντων όλων των κατοίκων της 'Ανατολής.

Ο Γενικός διευθυντής της 'Εφημερίδος,

ΣΤΕΦΑΝΟΣ ΞΕΝΟΣ.

No. 8.

2, Ingram Court, Fenchurch Street,
London, 12 March, 1859.

S. Xenos, Esq.,
Greek and Oriental Steam Navigation Company.

Dear Sir,

On Monday I shall, I hope, be able to be with you, and, to prevent any after-question arising, I think it but right to state the terms as I understand them. I am to be paid £350 per annum, and if the business progresses, and I give satisfaction, then an advance; I am also to have leave to fix any vessels belonging to my relations or friends, so long as I do not neglect your interests; and to conform to your instructions in all things. Trusting we may have great success,

I am, dear Sir,
Yours truly,
A. CARNEGIE.

No. 9.

Messrs. Overend, Gurney, and Co.,
    Lombard Street.

London, 23 May, 1862.

Gentlemen,

I have to request that you will advance to me the sum of seventeen hundred pounds for three months, by the discount of Messrs. Carr and Hoare's drafts on me for £1700, as below; and in consideration of your agreeing to do so, I give you a charge, and hereby charge my house and land at Petersham, which I have just bought, with the due payment of the said bills, and agree to execute any further document or deed that you may require, either by way of mortgage or otherwise, to complete your security on the said property; and in default of the due payment thereof, at maturity of the said acceptances, I hereby authorize you to sell the said securities as you may deem fit, and engage to pay any deficiency.

STEFANOS XENOS.

Dfts.—Messrs. Carr and Hoare on S. Xenos, Esq.,
    Greek and Oriental Company.
    £850 } dated 23rd May, 1862.
      850 } due 26th Aug., 1862.
    ─────
    £1700

No. 10.

29th of July, 1862.

Dear Mr. Carr,

I am now prepared to pay you the £1700 that Messrs. Overend, Gurney, and Co. have kindly advanced me to complete the purchase of my house, Petersham Lodge. This with some furniture, as I told you before, I bought from my private resources. My books and other furniture are what I have collected the last 16 years I have resided in England, and they have nothing whatever to do with the estates or properties of the Greek and Oriental Steam Navigation Company. This day I have none other creditor whatever, except Messrs. Overend, Gurney, and Co., so if those gentlemen will give me a letter that they have no claim on my house, contents, and furniture, as indeed they have not, I intend to settle all in the name of my wife and children, so that if death occurs they may have some home and shelter. I should not have so disposed, but having been rejected, as you know, by the Commercial Life Insurance Company, also the Victoria Insurance, and having been told by the doctors that I am suffering from consumption, it is my duty as a father and husband to make some little provision for my family. I hope that Messrs. Overend, Gurney, and Co., who have ever treated me with such unexampled liberality and kindness, will consider the justice of my demand, and grant me my request.

Yours most truly,
STEFANOS XENOS.

No. 11.

Messrs. Overend, Gurney, and Co.,
    London.

London, 15th August, 1862.

Gentlemen,

I have to request that you will release my house at Petersham from all claim on your part, on my paying you the sum of seventeen hundred pounds on the 26th instant, in settlement of the acceptances of the Greek and Oriental Steam Navigation Company, due that day for the like amount, and in consideration of your agreeing to do so, I engage to follow your instructions in every respect in the conduct of the business of the Greek and Oriental Steam Navigation Company, and to transact no business on my own private account, except my literary, of any kind whatever, and to receive no letters on business without submitting them to be read by Mr. G. B. Carr, or by some one authorized by you; and further undertake to hand to you, as soon as received, all bills of lading and policies of insurance for grain or goods shipped on account of the Greek and Oriental Steam Navigation Company, and in all respects to conduct the affairs of the said company in such manner as may be agreeable to you; and for the said consideration, further agree to sign any document that you may require with reference to the said business of the Greek and Oriental Steam Navigation Company, when called upon by you to do so.

No. 12.

[*Private.*]

Stefanos Xenos, Esq.,
36, King William Street, E.C.

London, 19 *Aug.*, 1862.

Dear Sir,

We have received your letter of this date, and *in compliance with your request, we hereby release your house at Petersham, called "Petersham Lodge," from all claim on our parts*, on condition of your paying the bills for £1700, due 26th inst., drawn by Messrs. Carr and Hoare on the Greek and Oriental Steam Navigation Company, and of your carrying out what you undertake in your said letter.

We trust that the business of your company will now be conducted in a manner satisfactory to us, which you may depend will be to the advantage of all parties concerned.

We remain,
Yours truly,
OVEREND, GURNEY, AND CO.

No. 13.

Laurence Pountney Place,
17*th Dec.* 1862.

My dear Mr. Xenos,

I have great pleasure in sending you the enclosed letter from Overend, Gurney, and Co., which will set your mind at ease as to your house, but I hope you will live 50 years yet, and I do not think this is an exaggerated notion, if you will only *take reasonable care* of yourself.

With best wishes,

I remain,
Yours very sincerely,
G. B. CARR.

S. Xenos, Esq.

No. 14.

22, *Basinghall Street,*
9th May, 1861.

Dear Sir,

The bearer of this is my friend, Mr. Carr, who will represent me in superintending your affairs, in pursuance of the arrangement made at the time of the loan.

Mr. Carr will co-operate with you in the general management of the business, and I shall be guided entirely by his advice in giving you any money facilities; and I shall look to him to satisfy me as to the application of any sums that may be advanced.

Yours truly,

E. W. EDWARDS.

S. Xenos, Esq.,
Greek and Oriental Company.

No. 15.

The Greek and Oriental Steam Navigation Company,
London Street.

Lombard Street, 24 *May,* 1861.

Gentlemen,

Referring to our verbal arrangement, we now write you with reference to the establishment of the company on a proper footing, and beg to say that it is agreed that it will in future be worked by five large steamers, and five smaller ones, a list of which we annex, under the management of yourself and Mr. G. B. Carr; and it is understood that it is to continue its operations so long as it is conducted to the satisfaction of the said Mr. G. B. Carr, and the earnings of the ships are paid to us, but not to be less than two thousand pounds per month, in addition to the amount of interest accruing on the advances we have made to you.

In the event of our account being reduced so far that, on a fair estimate, our advance is fully covered by the said ships, then, on further payments being made, the said ships are each one to be released as the value thereof is paid off with interest, such value to be ascertained by two competent persons, one to be appointed by each party; but at no time are we to be left without a sufficient margin on our advance.

We shall thank you for a reply confirming the above.

I remain,
Yours faithfully,
OVEREND, GURNEY, AND Co.

The fleet to consist of the following ships :—

Scotia, Asia, Palikari, Mavrogordato, Leonidas,	large steamers.
Colletis, Colocotronis, Rigas Ferreos, George Olympius, Londos,	smaller steamers.

No. 16.

Addressed to ARISTIDES XENOS, ESQ.

Galatz, 15th November, 1861.

Dear Aristides,

I now proceed to answer your private letter of the 27th Oct., which I found awaiting me on my return from Saulina, on the 11th inst. And with feelings of both pain and satisfaction I do so; of pain, because, after all the years wherein you and your brother Stefanos have known me, and under all the circumstances, you should so far do me the injustice of hinting at and writing that you think some "diabolical dishonesty is at work." And more, that you should suppose that only personal gain should be my only moving power, and think that I am honest because I see my personal interest in being so.

Again, I read your letter with satisfaction because I appreciate a man telling me plainly what he thinks, and more also because I believe your letter was one of impulse, and written unpremeditated, and therefore not to be considered in all its force. In answering you, believe me, I have considered well all you have said, and have also thought well what is proper to answer under the circumstances; for, Aristides, if there is one iota of truth in your remarks and hints, you have only your duty to your brother and the company to perform, and that is, to remove me at once from Galatz as being altogether unworthy of the confidence placed in me heretofore, and because I have used that confidence to serve my own unjust ends.

To write you thus is a trial indeed, and a source of pain to me; but, Aristides, no other course is left open to me. I have always held out to both your brother and you that personal gain would and could never make me neither honest nor dishonest; and, I repeat, you fall short indeed of my idea of a man, if you think that honesty is to be bought with money. I make no boast nor brag when I say, once for all, I have within me that which tells me that to be honest, honourable, and independent is the privilege of every man, whatever his station in life may be. I have always acted and written so, although sometimes my motives have been misinterpreted; therefore, Aristides, I beg you to disabuse your mind on this matter at once and for ever, for, although the independence of my dear wife and child are to me a sacred matter, an honourable career and a clear conscience are to me infinitely dearer.

Now, as to the question of the Powerful, I can answer we loaded what we invoiced per this vessel for Leghorn, and this is proven by the vessel loading only 50 qrs. less this voyage than last, whilst then she had only 60 tons coals on board, and this voyage she had 160 tons at least. Last voyage she was loaded by the head, by filling the fore-hold; this voyage Davidson made a division in his fore-hold, and thus carried less grain in his fore-hold, and thus he kept the vessel by the stern; beside the 100 extra coals, she drew as much water last voyage as now, as, in fact, you will find on her arrival in London.

The Marco Bozzaris cargo cost more because the Tzamados and Bobolina had to load three to four miles from Giurgevo (and this never happened before in the Danube), and we had to pay the extra cartage. No one could have foreseen this, and why blame us? Are you not always telling us, " Buy cheap stuff in the up river?"

The present wheats, as you say, Theologos did telegraph you from Giurgevo, the 8000 qrs. would cost 31s. *ex oporto;* but then, Aristides, look at the telegram I sent you on the 16th Sept. from here, in which I tell you the wheats will cost 32s. so early as the 16th Sept., and you never answer, " Stop buying," but you answer, " Do not hesitate, pay 34s.;" and your letter adds, "pay 35s., only get the wheat good and about your sample." Now, all the wheats we have shipped are quite equal to our sample, clean, good, and heavy.

Now, why did Theologos say he could get 8000 qrs. at 31s.? Because, sir, he was at Giurgevo, and knew not a word about your having increased your original order for 10,000 qrs. for all October and November to 20,000 qrs., and then to 30,000 qrs. We could have got, no doubt, 8000 qrs. for October and November at 31s., for that small quantity could have been cleaned in Giurgevo; but then, Aristides, it is awfully different to get 30,000 qrs.

I declare and uphold that the business of the wheat this autumn is an honour to us; we have, with only 7 feet of water, brought down, cleaned splendidly, and shipped till now 15,000 qrs., and have other 8000 qrs. on the way here. We could have cleaned in Giurgevo the 8000 qrs., but we could never have cleaned properly there 30,000 qrs.; so we have brought the wheats here, landed them from one steamer, cleaned them, and shipped them on another. This alone, considering how well and thoroughly they have been cleaned, is equal to 1s. to 1s. 6d. per qr. Now, besides this, is the test, we could have sold, and were offered equal to 36s. and 37s. 6d. per qr., free on board here for these wheats. This is no story, but a fact well known here, and in itself speaks volumes. But enough of this, for it is of no use to speak in one's own favour.

You speak of your brother having sacrificed his own relations. I hope and believe he never sacrificed any one wilfully; but I know little of this, and cannot give any opinion. But let me assume that, if to-morrow Mr. Stefanos thinks his interests are prejudiced by being in my hands, then let him make such change as he will, and believe me, I will not consider myself sacrificed at all; for I have worked with good honest will, and firmly believe that, on the whole, the very best has been made of the opportunities.

And of this I am sure, I came here to discredit and confusion, short shipments, and universal unfriendliness, and these are all now reversed. For I prize the good name I hope I have, and the continuance of the friendship of your brother, which is not of yesterday, more than my present gains; although I should now be sorry to leave Galatz.

You speak of your father or Mr. Clenzo coming here. The latter I know not; but this I do know, that, as much as I respect your worthy father, he could never make the business here. From daylight till 8 P.M., often all night, we work, not in offices, but out here and there, sometimes Saulina Roads, other times in the interior, and more often loading and cleaning grain. And, Aristides, where would the Smyrna's case have been in these gentlemen's hands? Nowhere—lost. Now, whilst I write, I tell you, had I only permitted that case to have been neglected, without any one ever being wiser, the Austrians would gladly have given me £1000, and Theologos the same. And what makes me angry is this, that Mr. Theodore Xenos should get £60 and we nothing. What has he done? Arranged one witness to come here at the awful cost of £80, whilst he was a British subject, and could be summoned, and must have come at an expense of £1 1s. and travelling cost, about £15 more. But that's neither here nor there, and is for you to judge, not me. Whilst for fourteen days I wrote all day in the court and all night at home, searched out greasy old pilots, coaxed them and petted them—and nothing I hate worse than to be obliged to flatter that sort of men; but the case is gained, and I have upheld the opinion I gave, that I could win the case, and so am thoroughly satisfied. Aristides, if I cared or considered less of your business I would write less, so excuse the length I have gone to.

I only further add, why do you not send out some one to examine into the state of these things? or you may come—surely you can find the time—and then Stefanos would rest satisfied, and not live in a state of uncertainty and anxiety, as seems now to be the case.

You say, "Don't begin Stokar's games." Enough of this; I must be indeed a changed individual if I can have the spirit of that man in me.

I repeat, the business here is done with the greatest amount of attention we can bestow, with the most honest purpose in view, and with the greatest endeavours to promote the interests intrusted to us; and all we ask is a proper and thorough inquiry into the whole matter of these wheats (and all others), and you will find not 1 cent wrongly charged nor 1 cent wasted in expenses that could have been avoided.

I must tell you that Mr. Theologos most fully concurs in all I have written, and on his arrival in London will personally confirm my statement; for this letter comes as much from him as me.

Awaiting your reply, and trusting you will be satisfied with this letter,

I remain, yours very truly,

A. CARNEGIE.

No. 17.

London, 24th Dec., 1861.

S. Xenos, Esq.
Dear Sir,

The SS. Palikari having been ordered for survey by Government for the conveyance of stores from Woolwich to Bermuda, you will be good enough to place the steamer in the hands of Messrs. Thompson and Tweeddale, of 27, Birchin Lane (who tendered her), who will act as the brokers.

I remain,
Dear Sir,
. Yours most truly,
E. W. EDWARDS.

No. 18.

Dear Xenos,

I am sorry I did not see you on Tuesday. I now enclose letter which it was agreed between Mr. Carr and myself that I should write to you upon the subject of the Palikari.

Yours most truly,
E. W. EDWARDS.

I strongly recommend you to call *yourself* upon the Comptroller.

London, 2 Jan., 1862.

No. 19.

London, 2nd Jan., 1862.

Sir,

The Palikari.

The bearer of this is Mr. Stefanos Xenos, the owner of the above-named ship, of which I am mortgagee. It was unfortunate that, in consequence of illness, I was unable to communicate with Mr. Xenos at the time I authorized Messrs. Thompson and

Co. to tender the ship to the Government for the conveyance of goods to Bermuda, as I am sure, had I done so, all the unpleasantness that has occurred, and which I regret extremely, would not have occurred.

Mr. Stefanos Xenos is alone entitled to sign charter-parties; and I think, after this explanation, you will be induced to accept any of his ships which may hereafter be tendered by him, and which you may consider fitted for your service. Apologizing for being very unintentionally the means of giving trouble,

I am, Sir,

Your obedient servant,

E. W. EDWARDS.

Chas. Richards, Esq.,
    Comptroller, &c.,
        Admiralty.

No. 20.

The Greek and Oriental Steam Navigation Company.

19, *London Street, Fenchurch Street,*
London, E.C.,   186

To the Bank of London.

Sir,

I enclose Mr. Sidney Malettass's, of Smyrna, drafts on your bank for fifty thousand pounds, at six months' sight, as particularized to you in those letters of 20th April, which I thank you to accept to the debit of this company against securities herewith mortgaged of our steamer—

Palikari,	£16,000
Mavrogordato,	14,000;
Leonidas,	6,000;
Zaïmis,	7,000;
Colocotronis,	7,000;

also a letter from Messrs. O., G., and Co., undertaking to provide you with the fifty thousand at the maturity of the bills, in case we fail to do so.

No. 21.

The Greek and Oriental Steam Navigation Company.

19, *London Street, Fenchurch Street,*
London, E.C., 28 October, 1861.

Received from Stefanos Xenos, Esq., the following Policies :—

No. 95, Marco Bozzaris,			£9,000
,, 2,	,,	,,	1,500
,, 3,	,,	,,	4,500
,, 4,	,,	,,	3,000
			£18,000

This steamer having been lost, I promise to pay the money to him, and not to any one else, in due course.

M. E. MAVROGORDATO.

No. 22.

The Greek and Oriental Steam Navigation Company.

19, *London Street, Fenchurch Street,*
London, E.C., 28 October, 1861.

Gentlemen,

I acknowledge receipt of original protest covering the loss of the SS. Marco Bozzaris, which I will return after the entire settlement with the underwriters.

To the G. and O. S. N. Co.

I remain, Gentlemen,
Your obedient servant,
M. E. MAVROGORDATO.

No. 23.

Laurence Pountney Place,
Saturday, 28th Feb., 1863.

Dear Sir,

Your conduct is very annoying to all interested in the G. and O. Stm. Co., and destructive to yourself—in fact, suicidal, as regards your position and future prospects.

You are generously offered a release from a load of debt, with the simple condition of engaging, under a proper penalty, to afford every assistance to getting in the assets of the company, and avoiding everything offensive to our friends. If you assign all over at once to my firm, and give the guarantees I have named, your real position will never be known among your countrymen or other friends; whereas, if you are any longer obstinate, and determined upon your own destruction, Carr and Hoare will be under the necessity of striking a docket against you in a day or two, to protect their own interests, as well as the interest of those friends whom you have treated so badly, in return for all their liberality and kindness to you. I hope that you possess copies of your letters to them of the 21st Jan. and 28th May, 1861, and of the 19th Aug., 1862; if so, read them, and then, if you are not mad, you will at once fulfil your promises therein expressed, and endeavour to propitiate their friendly feelings towards you for the future.

I remain,
Yours faithfully,
G. B. CARR.

Stefanos Xenos, Esq.

No. 24.

73, New Bond Street,
Augt. 21, 1868.

Sir,

In reply to your favour, having reference to some papers and property contained in a case supposed to have been removed with other property from Petersham, after your sale there, by Camp, the carman, to his warehouse, I can only suggest again that you had better make an appointment to go over his premises, and see if you can point it out. He believes that he has nothing of the kind, and that you had away everything which he removed, and it may be regretted now that the appointment which you made a long time ago for that purpose was not carried out.

If you will now suggest a time, giving a day's notice, I will arrange with Camp to meet you at the warehouse, which is a the top of Portland Street, in the New Road, opposite Trinity Church.

The only property which came to my house were the pictures.

I am, Sir,

Your obliged and obedient servant,

WM. PHILLIPS.

Stefanos Xenos, Esq.

No. 25.

16, *Tokenhouse Yard, E.C.*
London, 15 *June*, 1861.

Re Lascaridi and Co.

Gentlemen,

We beg to enclose an account current between the above firm and your company, showing a balance against you, with interest to the 31st May last, of £19,720 10s. 6d., of which amount we have to request payment on behalf of Messrs. Lascaridi and Co.

We remain, yours faithfully,

COLEMAN, TURQUAND, YOUNGS, AND CO.

The Greek and Oriental Steam Navigation Company.

No. 26.

16, *Tokenhouse Yard, E.C.*
London, 15 *June*, 1861.

Lascaridi and Co.

Sir,

We beg to enclose you accounts between yourself and Messrs. Lascaridi and Co., namely—General account, showing balance to your debit of £2189 2s. 4d.; and freight account, balance also to your debit, £1440 7s. 8d. Both accounts have interest calculated to the 31st of May last.

We have to apply on behalf of Messrs. Lascaridi and Co. for payment of the above balances.

We remain, Sir,

Yours faithfully,

COLEMAN, TURQUAND, YOUNGS, AND CO.

Stephen Xenos, Esq.

No. 27.

16, *Tokenhouse Yard, E.C.*
London, 20 *June*, 1861.

Re Lascaridi and Co.

Sir,
We are in receipt of your favour of to-day's date. We think that any objection you may have to the accounts we furnished to you should be more specific than that which your letter indicates.

If you have any account against Messrs. Lascaridi and Co., we have to request that you will forward the same to us. Messrs. Lascaridi and Co. have nothing whatever to do with any account which your company may have with their correspondents at Constantinople, or with parties at Odessa, Beyrout, or Galatz, who may happen to be agents of Messrs. Lascaridi and Co. at those places.

Waiting the favour of your reply,

We are, Sir,

Yours faithfully,

COLEMAN, TURQUAND, YOUNGS, AND CO.

S. Xenos, Esq.
Greek and Oriental Steamship Company,
Mark Lane.

No. 28.

1 *Corn Exchange Chambers, Seething Lane,*
London, E.C., 7/9/61.

Stefanos Xenos, Esq.,
London.

Dear Sir,
We have your letter of 6th inst., and, in reply, have to state that, having sold the wheat by orders of Messrs. Overend Gurney, and Co., and with their approval, we cannot see any case for arbitration, and clearly no case between you and us.

We are, respectfully,

COVENTRY, SHEPPARD, AND CO.

No. 29.

The following conditions to be agreed on between Mr. G. B. CARR and S. XENOS, for the purpose of bringing out a steam company:—Mr. G. B. Carr to become a director of the new company, and take shares of the value of £5000 to £10,000 when paid up. The boats Powerful, Asia, Scotia, Palikari, and Mavrogordato, now Mr. G. B. Carr's property, to be purchased by the new company for £   , if so desired by Mr. Carr, within three months from this date. Mr. G. B. Carr sells the goodwill of the Levant and Black Sea line of steamships to this new company for Two thousand five hundred pounds; Mr. Carr to secure the moral influence of Messrs. Overend, Gurney, and Co., to assist the new company. The office and clerks of the L. and B. S. line of steamships to be taken over by new company, except Mr. Ross and Mr. Papandenopulo; and whatever dépôt of coals, store, warehouses, wharfs, materials, &c., office fittings, &c., to be taken over by new company at cost price, usual wear and tear being allowed. A brokerage of 2½ per cent. to be paid by Mr. Carr to A. Carnegie, in case of disposal of his steamers to the new company.

No. 30.

COPY OF DEED OF ASSIGNMENT.

This Indenture, made the thirty-first day of March, One thousand eight hundred and sixty-three, BETWEEN STEFANOS XENOS, of King William Street, in the City of London, Merchant and Shipowner, of the first part, and EDWARD WATKIN EDWARDS, of Basinghall Street, in the said City of London, Esquire, of the second part, and SAMUEL GURNEY, HENRY EDMUND GURNEY, DAVID WARD CHAPMAN, ARTHUR GEORGE CHAPMAN, and ROBERT BIRKBECK, all of Lombard Street, in the said City of London, Money Dealers, carrying on business under the style or firm of OVEREND, GURNEY, AND COMPANY, of the third part.

Whereas the said Stefanos Xenos, who for some time past has carried on the business of a merchant and shipowner, under the style or title of the Greek and Oriental Steam Navigation Company, has contracted and agreed with the said Edward Watkin Edwards for the absolute sale to him of the whole of the ships, property, and assets of such business, subject, nevertheless, to the debts and liabilities now subsisting in respect thereof, for the sum of Two thousand five hundred pounds. AND WHEREAS there have been extensive dealings and transactions between the said Stefanos Xenos and the said parties hereto of the third part, in the course of his said business, and upon the balance of account, if now made up, the said Stefanos Xenos would be indebted to them in a considerable sum of money, as appears by the account submitted to the said Edward Watkin Edwards; and the said parties hereto of the third part, being fully satisfied of the ability of the said Edward Watkin Edwards to discharge the several debts and liabilities of the said business, have agreed to join in these presents for the purpose of releasing the said Stofanos Xenos from all claims and demands in respect of such dealings and transactions, and have further agreed, at his request, to become sureties for the said Edward Watkin Edwards, in manner hereinafter appearing.

And whereas, in part performance of the said agreement, the said Stefanos Xenos hath, by bills of sale bearing even date herewith, in consideration of the several sums of money therein mentioned, to be paid by the said Edward Watkin Edwards, amounting in the whole to the sum of One thousand five hundred pounds (being part of the said sum of Two thousand five hundred pounds), duly transferred the whole of the ships or vessels, and parts of ships or vessels, belonging to him, and employed in his said business, together with all and singular the appendages and appurtenances whatsoever thereto belonging, or in anywise appertaining to the said Edward Watkin Edwards, or to his nominees, subject to the several registered mortgages thereof.

Now this Indenture witnesseth that, in further pursuance and performance of the said agreement, and in consideration of the sums of money, amounting in the aggregate to One thousand five hundred pounds, so paid by the said Edward Watkin Edwards, as hereinbefore is mentioned, and of the sum of One thousand pounds of lawful money to the said Stefanos Xenos in hand, well and truly paid by the said Edward Watkin Edwards, at or before the execution of these presents, the receipt of which said several sums of One thousand five hundred pounds and One thousand pounds, making together the sum of Two thousand five hundred pounds, he, the said Stefanos Xenos, doth hereby acknowledge, and of and from the same, and every part thereof, doth hereby acquit, release, and discharge the said Edward Watkin Edwards, his executors, administrators, and assigns, for ever by these presents. HE, the said Stefanos Xenos, DOTH, by these presents, bargain, sell, assign, and transfer unto the said Edward Watkin Edwards, his executors, administrators, and assigns, ALL and singular the debts, credits, monies, freights, books, office furniture, bills and securities for money, property, and effects, whatsoever and wheresover, of and belonging to him, the said Stefanos Xenos, in the trade or business of a merchant and shipowner, so as aforesaid carried on by him under the style or title of the Greek and Oriental Steam Navigation Company, together with the goodwill of the said business, and all other the property, right, title, interest, claim, and demand whatsoever, both at law and in equity, of him, the said Stefanos Xenos, therein or thereto, TOGETHER with all books, writings, deeds, bills, notes, and receipt papers and vouchers touching the same, or any part thereof; and together also with full power and authority to and for him, the said Edward Watkin Edwards, his executors, administrators, and assigns, at all times, in the name or names of the said Stefanos Xenos, his executors or administrators, or otherwise, to ask, demand, sue for, recover, and receive, and give effectual releases and discharges for the monies and premises hereby assigned, or intended so to be. TO HAVE, HOLD, RECEIVE, AND TAKE the said debts, credits, monies, freights, securities for money, goods, chattels, property, and effects, and all and singular other the premises hereinbefore assigned, or intended so to be, unto the said Edward Watkin Edwards, his executors, administrators, or assigns, for his and their own absolute use and benefit. AND the said Stefanos Xenos doth hereby, for himself, his heirs, executors, and administrators, covenant and declare and agree with and to the said Edward Watkin Edwards, his executors, administrators, and assigns, in manner following, that is to say, that he, the said Stefanos Xenos, hath not at any time heretofore made, done, executed, or knowingly suffered any act, deed, matter, or thing whatsoever, by means whereof he is prevented from effectually assigning the aforesaid ships, vessels, freights, debts, credits, goods, chattels, effects, and premises, subject and in manner aforesaid, and according to the true intent and meaning of these presents. AND FURTHER, that he, the said Stefanos Xenos, his executors and administrators, shall and will, at the request, costs, and charges of the said Edward

Watkin Edwards, his executors, administrators, or assigns, make, do, and execute, or cause and procure to be made, done, and executed, all such further acts, deeds, and assurances whatsoever for the more effectually assigning and assuring the said premises unto the said Edward Watkin Edwards, his executors, administrators, and assigns, in manner aforesaid, as by the said Edward Watkin Edwards, his executors, administrators, or assigns, may be reasonably required. AND the said Edward Watkin Edwards, for himself, his heirs, executors, and administrators, and likewise the said parties hereto of the third part, for themselves, their heirs, executors, and administrators, as his sureties, do, and each of them doth, hereby covenant and agree with and to the said Stefanos Xenos, his executors and administrators, that he, the said Edward Watkin Edwards, his executors, administrators, or assigns, shall and will well and truly and punctually pay and discharge the several debts and liabilities of him, the said Stefanos Xenos, in respect of the aforesaid trade or business, except any debts or liabilities which the said Stefanos Xenos may have in any manner concealed, and shall and will from time to time, and at all times hereafter, save harmless and keep indemnified the said Stefanos Xenos, his executors and administrators, from all claims and demands, costs, charges, and expenses in respect of the same, and every part thereof.

And this Indenture further witnesseth that, in pursuance of the said recited agreement, and in consideration of the assignment hereinbefore contained on the part of the said Stefanos Xenos, they, the said parties hereto of the third part, do, and each and every of them doth, by these presents, absolutely acquit, release, exonerate, and for ever discharge the said Stefanos Xenos, his heirs, executors, and administrators, of and from the said claims and demands in respect of the before-recited dealings, and all manner of actions and suits, cause and causes of actions and suits, claims, and demands whatsoever, which they the said parties hereto of the third part, or any or either of them, now have, or which they or any or either of them, their heirs, executors, or administrators, might, but for these presents, have had upon or against the said Stefanos Xenos, his heirs, executors, or administrators, under or by virtue of such dealings and transactions as aforesaid, or otherwise howsoever.

In witness whereof the said parties to these presents have

hereunto set their hands and seals, the day and year first above written.

(Signed)

Signed, sealed, and delivered by the above-named Stefanos Xenos, in the presence of William J. Crossfield, clerk to Messrs. Thomas & Hollams, Solicitors, Mincing Lane, London. } STEFANOS XENOS.

Signed, sealed, and delivered by the above-named Edward Watkin Edwards, Henry Edmund Gurney, David Ward Chapman, Arthur George Chapman, and Robert Birkbeck, in the presence of Charles Edward Jones, 2, St. Mildred's Court. } E. W. EDWARDS.
SAML. GURNEY.
H. E. GURNEY.
D. WARD CHAPMAN.
ARTHUR G. CHAPMAN.
ROBT. BIRKBECK.

Signed, sealed, and delivered by the said Samuel Gurney, in the presence of Richard Fry, 136, Fenchurch Street.

No. 31.

Memorandum of the terms agreed for transfer by Mr. S. XENOS of the business and property of the Greek and Oriental Steam Navigation Company to Messrs. OVEREND, GURNEY, AND COMPANY.

1. Mr. Xenos to receive Two thousand five hundred pounds cash forthwith.
2. Mr. Xenos to be at liberty to send two servants and furniture, &c., not exceeding fifty tons measurement, by the Asia, free of freight, to Athens, or to the nearest port to that city at which she may call on her next outward voyage.
3. The acceptance for £2500 held by Mr. Carr to be paid out of the surplus proceeds of the wheat in the hands of Messrs. Alexander Bell and Company; but, if such proceeds are not sufficient for the purpose, Mr. Xenos not to be liable to pay the bill. Mr. Carr to have such control over the sale of the

wheat as Mr. Xenos now has. Any surplus in respect of the wheat to belong to Mr. Xenos.

4. Mr. Xenos to be indemnified against all the debts and liabilities of the Greek and Oriental Steam Navigation Company, but the indemnity not to extend to any debts or liabilities (if any) which he may have concealed. Mr. Xenos to be released from all liabilities to Messrs. Overend, Gurney, and Company, or to their nominees.

5. Mr. Xenos to give a bond for £2500, conditioned for his rendering all reasonable assistance in winding up the affairs of the company, and realizing the assets, &c., and doing nothing to injure or annoy the company or those interested in it; and also binding him that he has not, since 1st of January last, received any freights or moneys, or withheld any bills of lading or documents of the company not disclosed by the books.

6. Mr. Xenos forthwith to transfer the ships, and to give up all books, papers, and office furniture, and to give a power of attorney to Mr. Carr to receive all freights, &c., and to sue in his name.

OVEREND, GURNEY, AND CO.

Dated 24th March, 1863.

No. 32.

Know all Men by these Presents, that STEFANOS XENOS, of King William Street, in the City of London, Merchant and Shipowner, lately carrying on business under the name, style, or title of the Greek and Oriental Steam Navigation Company, for divers good and valuable causes and considerations him thereunto moving, doth by these presents irrevocably make, constitute, and appoint GEORGE BOWNESS CARR, of Laurence Pountney Place, in the said City of London, Merchant, the true and lawful attorney of him, the said Stefanos Xenos, and in his name, or in the name of the said company or otherwise, to ask, demand, recover, and receive of and from all and every the persons or person liable to pay the same, all and every sum and sums of money whatsoever now due, or hereafter to become due to him, the said Stefanos Xenos, in respect of his aforesaid business; and upon receipt of any such sum or sums of money, to give, sign, and execute good and sufficient receipts, releases, and discharges for the same; and also to sign, make, endorse, or accept any promissory note or notes, or bill or bills of exchange; and also to sign, seal, execute, and deliver all such deeds, releases, or assurances in any way relating to the said business, as he may think fit; and also to commence and prosecute any actions, suits, or other proceedings at law or in equity against any person or persons in respect of any sum or sums of money due and owing, or to become due and owing, as aforesaid, and in respect of any claims and demands whatsoever of him, the said Stefanos Xenos, in his aforesaid business; and to appear to or defend any actions, suits, or other proceedings to be commenced or prosecuted against the said Stefanos Xenos in respect of his aforesaid business, to proceed to judgment and execution, or become nonsuit, or suffer judgment to go by default, in such actions, suits, or other proceedings, as to the said George Bowness Carr shall seem most expedient; and for all or any of the purposes aforesaid, to use the name of the said Stefanos Xenos, and one or more attorney or attorneys under him, the said George Bowness Carr, to nominate and appoint, and any such nomination at pleasure to revoke; and also to do, perform, and execute all such other acts, matters, and things whatsoever as shall or may be requisite or necessary in and about the premises, and generally to act in the management of the affairs and business of the said Stefanos Xenos, so carried on by him as aforesaid, in such manner as the said George Bowness Carr shall think fit, as fully and effectually, in all respects, as the said Stefanos Xenos could himself have done if personally present.

No. 33.

London, 31st *March*, 1863.

Mr. S. Xenos,

In consideration of your, at my request, signing the Power of Attorney in my favour already prepared, and also signing the other documents which have been prepared for transfer of the business and property of the Greek and Oriental Steam Navigation Company, I hereby agree not to enforce payment from you, or any other party thereto, of the bill of exchange for £2500 accepted by you and held by me, but to look solely to the surplus value (if any) of the wheat belonging to you, in the hands of Messrs. Alexr. Bell and Co., for payment of the said bill, with interest thereon.

Yours faithfully,
(Signed)   G. B. CARR.

No. 34.

London, to Wit—I, STEFANOS XENOS, of King William Street, in the City of London, Shipowner, do solemnly and sincerely declare that I have not concealed from Messrs. Overend, Gurney, and Company, or from Mr. George Bowness Carr, any debts or liabilities incurred by me under the firm of the Greek and Oriental Steam Navigation Company; and that, so far as such debts and liabilities are known to me (except disputed claims), they are disclosed by the books of account of the said company, or by contracts in the office of the company. And I make this solemn declaration conscientiously believing the same to be true, and by virtue of the provisions of an Act made and passed in the fifth and sixth years of the reign of His late Majesty King William the Fourth, intituled " An Act to repeal an Act of the present Session of Parliament, intituled ' An Act for the more effectual Abolition of Oaths and Affirmations taken and made in various Departments of the State, and to substitute Declarations in lieu thereof, and for the more entire Suppression of voluntary and extrajudicial Oaths and Affidavits, and to make other Provisions for the Abolition of unnecessary Oaths.' "

STEFANOS XENOS.

No. 35.

Received of Mr. Stefanos Xenos the following documents on cargoes of coal:—

Charter party......	per Mali Marco.	
Receipt for £15 10s.	do.	do.
Two bills of Lading	do.	do.
Policy for £300....	do.	do.

Two Bills of Lading per Ligure.
Charter-party ....   do.
Policy for £700 ..   do.

Two Bills of Lading per Australia.
Charter-party ....   do.
Policy for £510 ..   do.

Two Bills of Lading per Cattina.
Charter-party ....   do.
Policy for £470 ..   do.

Policies of cargoes the B. of L. of which are already sent:—
2 per Ginsachard............ £700
1 per Wings of the Morning .. 320
2 per James Cruikshank...... 475

Charter-party...... per Unita.
Policy for £650 ....   do.

G. B. CARR.

No. 36.

The Greek and Oriental Steam Navigation Company.

36, *King William Street*,
London, E.C., 19*th May*, 1863.

Stefanos Xenos, Esq.,
    Petersham Lodge,
        Petersham, near Richmond.
Sir,
    It is my unpleasant duty to inform you that Mr. Lucas Ralli, of the Piræus, writes to the G. & O. S. N. Co. that he has no coals belonging to the company, having transferred about 845½ tons, by your orders, to some private account with you. I now learn from Mr. Ross that you directed him to pass to the credit of your private account the value of 270 tons, @ 26/- (being £351), for coals taken by the Mavrocordatos, when at that port, in February last year.

    I have also to inform you that Mr. F. Clenzo, of Smyrna, states that he paid you, on the 8th May last, when in London, the sum of £180 12s. 10d. on account of coals and advances per Parthian, which has not been passed through the company's books.

These several matters I must request you at once to set right, in the meantime

I remain,
Sir,
Your obedient servant,
For the Greek and Oriental Steam Nav. Co.,
G. B. CARR.

No. 37.

G. B. Carr, Esq.,
Laurence Pountney Place.

Lombard Street, 10th *March*, 1862.

Dear Sir,

With reference to the prospectus of the company enclosed herein, we are very much surprised to see Mr. Xenos's name down as a director, which is contrary to our understanding, even if it was likely to be a benefit to the G. & O., and we trust you will see that his name is withdrawn from it, and think he should not have gone into it without first consulting you on the subject.

We remain,
Yours faithfully,
OVEREND, GURNEY, AND CO.

No. 38.

KINGDOM OF GREECE.

ATHENS,
$\frac{20\ \text{Janvier},}{1\ \text{Février},}$ 1865.

MINISTRY OF HOME AFFAIRS.

Aux très-honorables James Wyld, membre du Parlement d'Angleterre, et Stéphanos Xénos, etc.

Messieurs,

En réponse à votre lettre en date du 5 Décembre, 1864, j'ai l'honneur de vous informer que quelque soit mon désir de contribuer à la consolidation de l'association formée entre des personnes si haut placées dans l'estime publique dans un but éminemment utile à la Grèce, quelque soit l'envie que j'éprouve de mettre ces personnes un moment plutôt en mesure de réaliser les deux beaux projets qu'elles ont conçus, il m'est impossible, à mon grand regret, de déposer entre leurs mains l'acte officiel que vous demandez par votre lettre, et qui engagerait si avant la responsabilité ministérielle.

Il est vrai que cet engagement n'aurait pour l'état aucune conséquence onéreuse, dans le cas où le Corps Législatif rejetterait les projets qui lui seront présentés aussitôt sa prochaine réunion ; mais il est également vrai que le rejet des projets, ou leur modification dans un sens qui n'aurait point votre assentiment, prouverait péremptoirement l'inutilité de l'engagement sollicité.

Mais tout en n'obtempérant pas directement à votre demande, je vous puis assurer de mon intention de présenter au Corps Législatif les deux projets de loi, tels qu'ils ont été rédigés et modifiés sur les observations du très-honorable Monsieur A——, si à cette époque je continue à occuper le poste de ministre compétent de représenter l'utilité des entreprises projetées, ainsi que les avantages réciproques des conditions auxquelles vous vous offrez pour en devenir les concessionnaires.

La Grèce, Messieurs, occupée du laborieux travail de sa régénération morale et matérielle, appelle tous les concours ; celui que lui offrent des personnes comme vous, placées au centre des grandes affaires, où affluent les capitaux du monde, lui serait trop précieux pour ne pas être accepté avec empressement par ses représentants.

Agréez, Messieurs, les assurances de ma considération très-distinguée.

A. COUMOUNDOUROS.

No. 39.

Messrs. Overend, Gurney, and Co. beg to inform the Greek and Oriental Steam Navigation Company that they have handed to Messrs. Coventry, Sheppard, and Co. the bills of lading for the wheats per Asia and Mavrogordatos, with instructions to land the same for them; and they have to request that the wheat per

Circassian, already landed, belonging to Messrs. Overend, Gurney, and Co., but standing in the name of Mr. S. Xenos, may be immediately transferred into their name.
Lombard Street,
14 *July*, 1862.

19, London Street, E.C.

No. 40.

The Greek and Oriental Steam Navigation Company.

19, *London Street, Fenchurch Street,*
London, E.C., July 4*th*, 1862.

Messrs. Theologos and Carnegie,
Galatz.

Dear Sirs,

Since our last, we are in receipt of your letters of 17th, 19th, and 20th ult., contents of which are duly noted.

We are sorry to say that we can find no sale at all for our wheats, and yet many of the Greek houses who have cargoes daily arriving, find a ready sale for theirs, at greatly improved prices. We have lately been much astonished at this, and having consulted with many other merchants, we find that not only is the quality so much better as to fetch four or five shillings more than any offer we can get for ours, *but the cost is invariably one or two shillings less than ours.* We learn from them that the reason of their wheats being so much cleaner than ours, is owing to the machinery employed in cleaning them; and as a machine has lately been invented which more effectually answers this purpose, we are now testing it here, and if we find it satisfactory, we shall send some of them out to you, with an engine and engineer to work them. This we are sure will soon pay for itself, and in the meantime we recommend you to be very cautious in your purchases, as it is yet unaccountable to us how our grain costs more than that of other merchants. This we wish you to understand is not mere hearsay, *but an actual fact to be shewn by the invoices.*

It has lately come to our notice that your Mr. Carnegie has been doing a considerable deal of business on his own account. This, we must remind you, we cannot allow, and your attention must be either *wholly* and *solely* devoted to the interests of the company, or the alternative.

Yours,

STEFANOS XENOS.

No. 41.

The Greek and Oriental Steam Navigation Company.

19, *London Street, Fenchurch Street,*
London, E.C., June 5th, 1862.

Messrs. Overend, Gurney, and Co.,
London.

Gentlemen,

Messrs. Coventry, Sheppard, and Co., the holders of my wheats for sums of money advanced to me at various times by you, have begun to sell, without any instructions from my part, at 2/. to 3/. per quarter less than what we could have got 4 weeks ago— viz., when the market was 3/. under what it now is—viz., 6/. less.

They tell me such is your decision. I protest against such proceedings, that will be most ruinous at least to me; and I beg to give you notice that I am now fully decided not to go any further with this business of the Greek and Oriental Steam Navigation Company, unless a written agreement is drawn up, in which I can see a fair prospect that I shall not have to work all my life (as I did until now) merely to fill other people's pockets. For your information, I tell you that I yesterday refused 42/. for 2000 qrs. from Mr. Lucas, ex ship—viz., nearly 1/6 more than what Messrs. Coventry, Sheppard, and Co. sold at the same day.

I believe the wheat sold by Messrs. Coventry and Co.—viz., 3000 qrs. ex Palikari, and 500 qrs. ex Colocotronis, is sold even without the consultation of Mr. Carr.

I remain, Gentlemen,
Yours truly,

STEFANOS XENOS,
For the Greek and Oriental Steam
Navigation Company.

No. 42.

Stefanos Xenos, Esq.,
    To G. B. Carr.
    For Sundries under Agreement dated 25 Apl. 1865.

June, 1865.	Charges.							
Westcott, Salary per Month,	£25	0	0					
Housden,	,,	,,		12	10	0		
Hammond,	,,	,,		10	0	0		
Twiddy,	,,	,,		5	0	0		
							£52	10	0
July, 1865.								
Westcott,	,,	,,		25	0	0		
Housden,	,,	,,		12	10	0		
Hammond,	,,	,,		10	0	0		
Twiddy,	,,	,,		5	0	0		
							£52	10	0
Augt.								
Westcott,	,,	,,		25	0	0		
Housden,	,,	,,		12	10	0		
Hammond,	,,	,,		10	0	0		
Twiddy,	,,	,,		5	0	0		
							£52	10	0
Sept. 8th.								
Westcott,	,,	,,		6	13	4		
Housden,	,,	,,		3	6	8		
Hammond,	,,	,,		2	13	4		
Twiddy,	,,	,,		1	6	8		
							£14	0	0
							£171	10	0

Proportion of Office Rent, June 24 / Sept. 8 .. 48 19 5
Do. cleaning Offices, ,, 3 6 0

£52 5 5

£223 15 5

2nd Oct., 1865.
Received check ℙ £223 15s. 5d.,
    G. B. CARR.

B B

No. 43.

Laurence Pountney Place,
9th October, 1865.

My dear Sir,

I beg to acknowledge the receipt of your acceptances to Theologos and Carnegie ℔ £1000 at 3 m/d, and £1000 at 6 m/d, from the 30th ult., also the Anglo-Greek Steam Navigation Company's acceptance to A. Iovanoff, at 4 m/d, ℔ £1500, making together £3500, being the consideration for the goodwill of the Levant and Black Sea line of steamers, and for their furniture, fittings, &c., at No. 9, Fenchurch St., as per our agreement.

I remain, yours very truly,
G. B. CARR.

S. Xenos, Esq.

No. 44.

9, *Fenchurch Street,*
25*th Oct.,* 1865.

G. B. Carr, Esq.

Dear Sir,

I hand you herewith two scrip certificates of shares in the Anglo-Greek Steam Navigation Company, Limited—viz., No. 42, for 100 shares, paid up to £1600; and No. 39, for 31 shares, paid up £496—as collateral security for my acceptances ℔ £1000 due 2nd January, and ℔ £100 due 2nd April; and if these acceptances are not paid at maturity, I hereby authorize you to dispose of the shares, for which I enclose transfers, the surplus to be paid over to me—or if any deficiency, I will pay the same to you with interest.

I remain, dear Sir,
Yours very truly,
STEFANOS XENOS.

No. 45. (*Vide* page 282.)

[*Copy.*]

7, Great Winchester Street,
8th May, 1865.

Messrs. Theologos and Carnegie,
Galatz and London.

Gentlemen,

In consideration of your subscribing for three hundred shares of the Anglo-Greek Steam Navigation and Trading Company, Limited, of £32 each, and which not more than £5 will be called for six months, I hereby appoint you agents of the company for the river Danube and all its ports, and guarantee this appointment with the directors, which shall be continued for three years. The terms to be not less than those allowed you when agents for the Levant and Black Sea Navigation Company.

(Signed)   STEFANOS XENOS.

No. 46.

Messrs. Overend, Gurney, and Co.

May 10th, 1861.

Gentlemen,

I received your very esteemed letter of yesterday, and, in answer, I agree on my part, and in behalf of the Greek and Oriental Steam Navigation Company, to every point of its contents; at the same time I take the opportunity to thank you most sincerely for your extreme kindness and liberality, assuring you that I shall spare neither trouble, pains, nor indeed any sacrifices, to make the said steamers answer, and thereby prove to you that I am not unworthy or ungrateful of the great and generous confidence you have shewn me. I cannot say more for the present, and would rather leave facts to speak for themselves.

Yours truly,

STEFANOS XENOS.

No. 47.

Messrs. Overend, Gurney, and Co.,
Lombard Street.

London, 26th March, 1862.

Gentlemen,

We have to request that you will give a guarantee to Messrs. Masterman and Co. for twenty-five thousand pounds, in respect of drafts of Messrs. J. Hamburger and Co., dated 2/14 March, 1862, at 90 days' sight, due 27 June, 1862, and in consideration of your agreeing to do so, we hereby charge the whole of the SS. ships, and the mortgages thereon lodged with you, with the due payment of the amount of the said guarantee, and undertake to pay you the said sum of twenty-five thousand pounds on the said 27 June, 1862, and in default thereof we hereby authorize you to sell the said securities as you may deem fit, and engage to pay any deficiency.

No. 48.

Iron Ship-Building Yard, Hebburn Quay,
Gateshead-on-Tyne, 16th Novr., 1860.

S. Xenos, Esq.,

My dear Sir,

I could not reply to your note yesterday, as I was out trying the screw yacht's engines. I am much pleased to inform you that everything went on to my full satisfaction. We got a speed of 11 knots, but I am sure she will go more when everything gets rubbed down. I enclose your certificate of official number, and both vessels will leave here in company for London to-morrow night, and I trust they will have a favourable run up. I regret the delay that has occurred very much, but the whole cause has been for the engines. If they had been got ready as promised to me, the vessels would have been with you by the middle of September. I, however, trust you will find them to your satisfaction, and have the opportunity of enjoying them long.

I am,

Yours faithfully,

ANDW. LESLIE.

No. 49.

Lombard Street, 13 March, 1861.
Dear Sir,
Would the Smyrna be of any use to you? She is offered for £6000, which is £2500 less than we got for her. Let us know to-morrow.

Yours truly,
WILLIAM BOIS.

G. B. Carr, Esq.,
Laurence Pountney Place, E.C.

No. 50.

Lombard Street, 24 Dec., 1861.
Dear Sir,
We have financed £25,000 on Masterman, due 27 March, on acct. of Greek and Oriental Steam Company, and endorse letter in respect of the same, which we shall be obliged by your getting signed and returned to us.

We remain,
Yours faithfully,
OVEREND, GURNEY, AND CO.

G. B. Carr, Esq.,
Laurence Pountney Place, E.C.

No. 51.

OVEREND, GURNEY, AND CO.
To the Editor of the TIMES.
Sir,
Allow me to send you the enclosed letter, this moment received from my son, Mr. D. Ward Chapman, on his hearing of Mr. Edwards's evidence in the distressing prosecution of Overend and Co. It speaks for itself, and I believe it to be perfectly true; and though it does not exempt him from his share of blame in the management of the house, it conveys a very different impression

from that which the ingenuity of counsel, speaking in the interest of his clients, was calculated to produce.

I would not say a word, at such a moment as the present, to add to the deep affliction of the Gurneys, in which nobody can participate more than myself; but on such a point, in which the highest moral principle is involved, it would be a dereliction of duty to keep silence.

I shall feel greatly obliged if you will insert this communication, as the subject is exciting such universal interest.

I remain, Sir,
Your obedient servant,
D. B. CHAPMAN.

Roehampton, Jan. 28, 9 A.M.

No. 51A.

Tours, Jan. 26.

My dear Father,

I have received your letter of the 25th inst., which I need not say has caused me the deepest pain it is possible for a man to feel. It is, of course, the business of Ballantine to enlist the sympathy of the court and the public for his clients, no matter at whose expense; but I can with a clear conscience, as if I were about to enter into the presence of my Maker, challenge the whole world to show, or even to state a single instance in which I have sacrificed the interest of Overend, Gurney, and Co., even in the smallest particular, to my own, or those of any of my friends. With regard to the transaction with Mr. Edwards, given in his evidence, at the time it took place I was in no sort of want of money, and it was offered by himself as a simple deposit at 5 per cent. interest. I fully acknowledge the extreme imprudence, and, as events have turned out, the impropriety of accepting the offer, but my private transactions with him never weighed one iota with me in my dealings for the house with him. The true state of the case was this— that after the commencement of the wild advances which resulted in the embarrassments and eventually the ruin of the firm, Edmund Gurney consulted with Mr. Edwards, who happened to be in the house upon some business connected with some bankrupt estate in which Overend, Gurney, and Co. were interested, and was so pleased with the sagacity which he thought was shown by Edwards, that he consulted him during a whole year upon many of what I may call the illegitimate advances which the firm had then made. At the end of the year he made himself, with the most casual consultation with me, a present of the £5000 mentioned in Edwards's evidence. Edwards, as he himself has since stated, and which in all subsequent negotiations with him has never been denied, said, in thanking Edmund Gurney, "I hope, Mr. Gurney, that this large payment is not intended as a farewell, but that I may hope that I may reckon on the continuance of my relations with the house." Edmund Gurney's reply was—" Friend Edwards, I do not see how we are to get on without thee." Not long after this J. H. Gurney came up, and was frightened at the amount of the involvements of the house. He found Edwards, who had up to that time no idea of anything like difficulties in the concern, confidentially engaged in the management of these accounts. He had several private conversations with him, which resulted in J. H. Gurney's telling him that the house was in a mess, but that there was nothing which it could not perfectly pull through, at the same time begging him to continue his supervision, and to keep him (J. H. Gurney) continually informed of the state of affairs. At the end of the year Edwards asked that his position should, in some way or other, be secured to him; and, indeed, I had considerable difficulty in persuading him not to ask for a partnership. Now, the position we were in was simply this: Here was a man who, not through me, notwithstanding my engagements with him on the subject (on what he was pleased to call my extreme reticence as regards the secrets of the house), had become possessed of sufficient knowledge to have forced us to have put up our shutters within twenty-four hours of his revelations, if he had chosen to make them. Now, I ask you what you would have done under such circumstances? Would you not have soothed him down in the best way you could, and made the best terms in your power? I will say myself, that any share that I might have had in the settlement with Edwards I would repeat again to-morrow. With regard to his statement that I have never accounted to him for the money, it is simply not true, as I paid him the interest regularly, and considerable amounts on account of the principal; and it was only at his own wish that I did not include him in the list of creditors under my deed, he preferring to run the chance of my honourably paying him in case I ever started afresh. Now a few words with regard to the great losses of the house. I neither began the accounts with Mr. Mare, which resulted in the Millwall Company, nor with Mr. Lever and the Atlantic Mail Company, nor with Mr. Xenos, all those three people being among the bitterest enemies I have, in consequence of their looking upon me as the person who stopped their depredations; but I confess to have taken the view that those accounts and their kindred ones were necessary to be kept out of the Bankruptcy Court by every means in our power, if we wished to retain the solid business which I always believed to be at the bottom of our embroilment. But it appears that I not only miscalculated the power of the house, but also the extent of the assistance which in time of need could be brought in. Let me state further a fact worthy of some consideration—that during the nine months in which the limited company was in existence, they managed to lose the sum of £1,300,000 by transactions entirely irrespective of the old firm; at least this is stated by Mr. Harding in his evidence. I will only add, that I was entirely unaware of Edwards having received any commission from any person with whom he was dealing on behalf of the firm. I hope this explanation may be satisfactory to you; but, whether it is so or no, my conscience accuses me keenly of many most culpable follies in my private capacity, but of disloyalty to the house never.

Believe me, your affectionate son,

D. WARD CHAPMAN.

No. 52.

To the Editor of the TIMES.

Sir,

The remarks in your columns to-day upon this case induce me to ask you for permission to explain a very great misconception which seems to have arisen from the reference made to my name by Mr. Edwards in his evidence before the Lord Mayor.

My connexion with Messrs. Overend, Gurney, and Co. commenced in the early part of 1861, when they requested me to take the superintendence of the works then in progress at Millwall, the property of Mr. C. J. Mare. I accepted that appointment at a fixed yearly salary of £1500. I continued as such superintendent until the formation of a company, in 1863, to purchase and work the establishment; the buildings and machinery (the erection of which was commenced by Mr. Mare) being then completed.

I had no further or other employment from Messrs. Overend, Gurney, and Co., except in connexion with those works.

Although I certainly had frequent transactions in business with Mr. Edwards after his connexion with Messrs. Overend, Gurney, and Co. had ceased, I never was a member of any firm called "Edwards and O'Beirne," nor did any such firm exist, to my knowledge.

No advances were ever applied for to Messrs. Overend, Gurney, and Co. by any firm or co-partnership of which I was a member, nor was I ever engaged in any pursuits which required such advances.

I am, Sir,

Your very obedient servant,

J. LYSTER O'BEIRNE.

36, *Sackville Street, Piccadilly, Jan.* 28.

www.ingramcontent.com/pod-product-compliance
Lightning Source LLC
Chambersburg PA
CBHW032031220426
43664CB00006B/434